INTRODUCTION TO NUTRITION
AND HEALTH RESEARCH

INTRODUCTION TO NUTRITION AND HEALTH RESEARCH

by

Eunsook T. Koh

and

Willis L. Owen

University of Oklahoma Health Sciences Center

KLUWER ACADEMIC PUBLISHERS
Boston / Dordrecht / London

Distributors for North, Central and South America:
Kluwer Academic Publishers
101 Philip Drive
Assinippi Park
Norwell, Massachusetts 02061 USA
Telephone (781) 871-6600
Fax (781) 681-9045
E-Mail <kluwer@wkap.com>

Distributors for all other countries:
Kluwer Academic Publishers Group
Distribution Centre
Post Office Box 322
3300 AH Dordrecht, THE NETHERLANDS
Telephone 31 78 6392 392
Fax 31 78 6546 474
E-Mail <services@wkap.nl>

 Electronic Services <http://www.wkap.nl>

Library of Congress Cataloging-in-Publication Data
Koh, Eunsook T.
    Introduction to nutrition and health research / by Eunsook T. Koh and Willis L. Owen.
        p. ; cm.
    Includes bibliographical references and index.
    ISBN 0-7923-7983-7 (alk. paper)
    1. Health—Research—Methodology. 2. Nutrition—Research—Statistical methods. I.
  Owen, Willis L. II. Title.
  [DNLM: 1. Nutrition. 2. Research—methods. 3. Research Design. 4.
  Statistics—methods. 5. Writing. QU 145 K787i 2000]
  RA645.N87 K64 2000
  610'.7'2—dc21

                                00-061056

Printed on acid-free paper.
Printed in the United States of America

The Publisher offers discounts on this book for course use and bulk purchases. For further information, send email to <joanne.tracy@wkap.com>.

CONTENTS

Preface   xiii

Acknowledgements   xv

Part I Overview of the Research Process   1

Chapter 1. Introduction to Research in Nutrition and Health   2

Definition of Research	2
Types of Research	3
Basic and Applied Research	3
Field versus Laboratory Research	3
Qualitative, Descriptive and Experimental Research	4
Unscientific and Scientific Methods of Problem Solving	5
The Rationalistic Method	6
The Empirical Method	6
The Scientific Method of Problem Solving	7
Summary	10
References	11

Chapter 2. Research Problem and Literature Review   13

Introduction	13
Identifying the Research Problem	13
Using Inductive and Deductive Reasoning	14
Purposes of the Literature Review	16
Primary and Secondary References	16
Reading and Recording the Literature	19
Critically Evaluating the Literature	21
Writing the Literature Review	22
Summary	23
References	25

Chapter 3. Framing a Research Problem: Hypotheses, Purposes, Objectives, and Questions   27

Choosing the Title   27
Writing the Introduction   29
Framing a Research Problem   33
Stating the Research Problem   34
Identifying the Variables   34
Hypotheses   36
Operational Definitions   37
Outlining Basic Assumptions, Limitations, Delimitations   37
Summary   39
References   40

Chapter 4. Writing Method Sections   41

Planning the Work   41
Describing Subjects   42
Subjects Sources and Selections   44
Diet and Other Interventions   48
Describing Instruments   49
Describing Procedures   49
Describing Design and Analysis   50
Statistical Analysis   50
Summary   54
References   55

Chapter 5. Ethical Issues in Research and Scholarship   57

Introduction   57
Misconduct in Science   57
Plagiarism   58
Fabrication and Falsification   59
Misleading Authorship   60
Ethical Issues Regarding Copyright   62
Protecting Human Subjects   63
Components of an Informed Consent   66
Guidelines for Using Human Subjects in Writing
       Proposals to the National Institutes of Health   67
Protecting Animal Subjects   68

Guidelines for Using Animals in Research Proposals to
      the National Institutes of Health   68
Institutional Animal Care and Use Committee   69
Summary   69
References   70

Part II. Statistical and Measurement Concepts in Research

73

Chapter 6. Statistical Concepts

74

Introduction   74
Applications of Statistics in Research   75
Levels of Measurements   78
Descriptive and Inferential Statistics   79
Sample Selection   80
Central Tendency   82
Two Categories of Statistical Tests: Parametric and
      Nonparametric Statistics   85
The Standard Normal Distribution   86
Application of the Normal Curve   90
The Normal Distribution as a Probability Curve   92
Hypothesis Testing   93
Null and Alternative Hypotheses   94
Two-Tailed versus One-Tailed Hypothesis Tests   94
Alpha (Type I error)   94
Beta (Type II error)   95
Meaningfulness (Effect Size)   96
Determining Required Sample Size   97
Summary   98
References   101

Chapter 7. Relationships Among Variables

103

Introduction: Correlation and Regression   103
Types of Relationships Between Two Variables   104
Applications of Correlation and Regression in Research   105
How Correlation Research Investigates   107
Correlation Coefficient   111
What the Coefficient of Correlation Means   112

Using Correlation for Prediction (Regression)   115
Interpreting Meaningfulness of r   118
Standard Error of Estimate   118
Partial Correlation   121
Multiple Regression Prediction Equation   121
Summary   123
References   125

Chapter 8. Differences Among Groups   127

Introduction   127
How Statistics Test Differences   127
Three Types of Student's t Test   128
Estimating Meaningfulness of Treatments   131
t Tests and Power in Research   135
Analysis of Variance : One way of ANOVA   136
Characteristics of the F Distribution   137
Follow-up Testing   140
Factorial Analysis : Two-Way of ANOVA   141
Analysis of Covariance   151
Summary   152
References   153

Chapter 9. Nonparametric Statistics   155

Introduction   155
Chi-square   156
Spearman Rank Correlation Coefficient   162
Mann-Whitney-Wilcoxon Test   165
Wilcoxon Matched-Pair Signed Ranks Test   165
Kruskal-Wallis ANOVA   166
Friedman's ANOVA   166
Summary   166
References   168

Chapter 10. Measuring Research Variables   169

Introduction   169
Comparisons of Scales: Practical Considerations and Statistics   169
Reliability   171
Precision and Accuracy   173

Validity   175
Measurement Errors in Dietary Assessment   177
Assessment and Control of Measurement Errors   179
Error Sources   180
Precision in Dietary Assessment   181
Validity in Dietary Assessment Methods   183
Use of Biochemical Markers to Validate Dietary Data   186
Anthropometric Assessment   188
Clinical Assessment   189
Summary   189
References   191

Part III Various Types of Research

195

Chapter 11. Experimental and Quasi-Experimental Research

196

Introduction   196
Internal and External Validity   197
Controlling Threats to Internal Validity   204
Controlling Threats to External Validity   206
Pre-Experimental Designs   207
True Experimental Designs   207
Randomized-Groups Design   208
Pretest-Posttest Randomized-Groups Design   210
Factorial Design   210
Solomon Four-Group Design   211
Quasi-Experimental Designs   213
Summary   215
References   216

Chapter 12. Descriptive Research and Qualitative Research

219

Definition of Descriptive Study   219
Definition of Qualitative Research   219
Survey Research   220
Questionnaires   220
Personal Interview   226
The Delphi Method   228
The Normative Survey   228

Nutrition Survey   228
Dietary Intake   229
Epidemiologic Descriptive Survey   232
Epidemiologic Approaches to Diet and Disease   232
Correlation Studies   233
Special Exposure Groups   234
Migrant Studies and Secular Trends   234
Case-Control studies   235
Cohort Studies   235
Controlled Trials   236
Cross Sectional Study   237
Meta-Analysis and Pooled Analysis   238
Qualitative Research   239
Summary   242
References   244

Part IV. Writing the Research Proposal and Results

249

Chapter 13. Results, Discussion, and Abstract

250

Introduction   250
Use of Tables in Data Presentation   251
Use of Graphs in Data Presentation   254
The Histogram   256
The Bar Graph   257
The Pie Chart   259
The Scatter Diagram   260
Photographs   263
Uniform Requirements for Manuscripts   265
Discussion   267
Abstract   269
Report of Qualitative Study   272
Summary   272
References   274

Chapter 14. Publications

277

Research Report   277
Thesis or Dissertation Writing Guidelines   277
Limitations of Chapter Style   278

Preliminaries   279
Body of the Thesis or Dissertation   283
Presentations and Manuscripts   288
Oral Presentations   288
Poster Presentations   290
Manuscripts   291
Author's Guidelines: 292
Submitting the Article   292
Summary   293
References   295

Chapter 15. Writing the Research Proposal   299

Introduction   299
Identifying the Research Problem   299
Choosing the Topic   300
Formulating the Hypotheses   300
Developing the Research Protocol   301
Abstract   302
Specific Aims   302
Background and Significance   303
Preliminary Studies/Progress Report   303
Research Design and Methods   304
Why Grant Proposal Fail   305
Suggestions for Grant Writing   305
How to Prepare a Strong Proposal   306
Summary   307
References   308

Part V. Using Computers in Research   309

Chapter 16. Using Computers   310

Introduction   310
Hardware   310
Software   312
Data Entry Software   312
Graphical Software   313
Statistical Software   313
Data Analysis Examples   314
Summary Statistics   315

A simple Histogram   320
A *t* test for independent groups   321
Analysis of a Contingency Table   323
Calculation of Correlation Coefficients   326
Simple Linear Regression   330
Multiple Regression   331
One Way Analysis of Variance without Repeated Measures   333
Two Factor Analysis of Variance for Independent Groups   335
Single Factor Analysis of Variance with Repeated Measures   336
Two Factor Analysis of Variance with Repeated Measures   339
Summary   341
References   342

Appendix   343

A.   Statistical Tables   343

Subject Index   358

Preface

A course on dietetics, nutrition and health research methods is generally one of the first courses both theses and non-theses students should take. Nonetheless, during the several years that we have taught the course "Research Methods in Dietetics, Nutrition and Health," we have not found any textbook on this critical topic. This book will hopefully fill the gap in dietetics, nutrition, and health educational literature.

This book is based on the research methods its authors have used to teach master's degree level introductory courses on dietetics, nutrition and health research methods as well as the feedback from the students in those courses. The primary objectives of the authors' courses are to teach students how to conduct their own research and how to understand the research of others. This book shares the same objectives and strives to be comprehensive yet still practical and understandable. This book is designed to help students create their own research projects and write their masters' theses as well as to teach students who will not write masters' theses how to analyze and interpret research articles. The book includes many refereed nutrition and medical journal articles so that students can learn how researchers identified research problems; how they framed the problems; how they planned and conducted research to address the problem; and how they reported, interpreted, and implemented their findings through real articles.

Doctoral students and beginning researchers who want an overview of the research process should find this book helpful. However, the procedures and techniques specific to a certain type of research in a specialized area are not covered in this book.

The growth of our professions depends upon being active producers and/or consumers of research. Researchers, teachers, technicians, and counselors need to understand the research process. I believe this book provides the necessary information for both the consumer and the producer of research.

Part I (Chapters 1-5) of the book provides a complete overview of the research process: developing a research topic, reviewing current literature on the topic, selecting a research approach, writing a prospectus, and analyzing the ethical issues in research and scholarship.

Part II (Chapters 6-10) discusses statistical and measurement concepts in research such as descriptive versus inferential statistics, parametric versus non-parametric procedures, and measurement issues associated with dependent variables. Part II is designed for students with no background in statistics and limited backgrounds in mathematics.

Part III (Chapters 11 and 12) explores the various types of research: experimental, quasi-experimental, descriptive, and qualitative research. This section also explains the various types of experimental designs.

Part IV (Chapters 13-15) explains how to present research results via oral and poster presentations, and how to write theses, dissertations, journal articles and research proposals.

Part V (Chapter 16) highlights the use of computers in research and demonstrates the various ways of data analysis.

Acknowledgments

This book is the product of many minds. Great appreciation is extended to Dr. Richard G. Allison, Executive Officer of the American Society for Nutritional Sciences for his supports and encouragement for me to complete this book.

Professor Judith C. Wilkerson and Mrs. Annette Moss have provided invaluable comments that have greatly improved the book. We have also striven to incorporate the astute feedback from our "Research Methods in Dietetics, Nutrition and Health" students on how this subject should be taught.

Thanks must be expressed to faculty and staff members in the Department and College who helped us to complete the book manuscript, especially, Gina R. Vile for her invaluable efforts in preparing the text and illustrations; and Gaylon E. Bright for his computer skills. The authors are very grateful to staff members, especially, Tony B. Cable and Michael Sweezy, in the Printing Services, Health Sciences Center, for their art works. This text would not have been possible without their contribution.

Finally, the authors wish to express their sincere thanks to Joanne Tracy, the Editor-in-Bio-sciences at Kluwer Academic Publishers, who gave essential support and counsel during the production of the book, and the editorial assistant, Mary Panarelli, for her technical advice and supports.

Eunsook T. Koh

Willis L. Owen

Needless to say, countless hours were expended by the co-author, Dr Willis L. Owen, to write Chapter 16 and to review all other chapters. Without his careful evaluation the present book would not have been possible.

Eunsook T. Koh

PART I

======================================

OVERVIEW OF THE RESEARCH PROCESS

This part provides a complete overview of the research process: developing a research topic, reviewing current literature on the topic, selecting a research approach, writing a prospectus, and analyzing the ethical issues in research and scholarship.

Chapter 1 defines and reviews the various types of research in nutrition and health. We discuss the steps in the scientific method: developing the problem, formulating hypotheses, gathering the data, and analyzing and interpreting results.

Chapter 2 suggests ways of developing a problem and using the literature to clarify the research problem, specify hypotheses, and develop the methodology. We also propose a system for searching, reading, analyzing synthesizing, and writing the review of literature.

Chapter 3 provides an overview of the study: statement of the purpose or problem to be addressed, title, introduction, hypothesis, delimitations, limitations, and significance of the study.

Chapter 4 covers methodology, or how to do the research. Included are the topics of subject selection, instrumentation of apparatus, procedures, and design and analysis.

Chapter 5 discusses ethical issues in research and scholarship. Ethics and procedures in the use of human and animal subject are also discussed.

CHAPTER 1

==

INTRODUCTION
TO RESEARCH IN NUTRITION AND HEALTH

Definition of Research

Research is simply a way of solving problems, and it is defined as planned studious and critical inquiry and examination aimed at the discovery and interpretation of new knowledge or expansion on a topic or idea. Quality assurance, surveys, new product evaluation, and case report are all examples of research. The purpose of these activities is to document "what works and what does not work," which answers particular research questions. Useful research studies may be done to substantiate other previous investigations. Studies of diet and disease relationship and studies of various nutrition therapies are absolutely indispensable in substantiating the value of nutrition in the health care environment.

In an editorial on research in the New England Journal of Medicine, Dr. Kahn stated:
"The first step in picking a research project is to understand what makes research good."

A good research project should be well planned, use up-to date technology, carefully analyze and accurately report the data, and appropriately deal with the ethical considerations associated with animals and humans when they are involved. An outstanding research project asks important questions, yields truly new knowledge, leads to new ways of thinking, and lays the foundation for other research in the field. Although these are major propositions for the new researcher, a good research project can be developed by keeping these points in mind when preparing and conducting any research endeavor, regardless of its size. Research in dietetics and nutrition is a dynamic and ongoing process. The answers will never be completely known to the myriad of potential research questions that are asked by the members of the nutritional health care team --- nutritionists, dietitians, physicians, nurses, biochemists, physiologists, and others. The key to answering some of these questions is for researchers to plan, organize, conduct, and communicate research findings.

Types of Research

Basic and Applied Research

Basic research is performed without a specific application purpose in mind. Rather, it is done primarily for the sake of knowledge alone. In contrast, applied research is done with a specific question or application in mind. Consider a nutrition educator who wishes to know if a six-week unit on nutrition education will change dietary habits and attitudes in high school students. The researcher wishes to address a specific question. In comparison, a basic nutrition researcher designs an experiment to find the role of insulin in fatty acid synthesis in the rat model. This somewhat vague approach demonstrates basic research. Although it is typically viewed that basic research is done with no specific application purpose in mind, one might question whether or not this is actually so. By observing what does actually happen, more links between observations, theories, and hunches may be developed. The results of this type of research may have no practical application. The majority of this research is done in highly controlled experimental laboratory settings, often with animal subjects. Selected variables have to be manipulated for maximum control.

It is important to realize that both types of research may exist on two ends of a continuum with neither one necessarily at the absolute end. For example, basic nutrition research uses the laboratory as the setting, frequently uses animals as subjects, carefully controls conditions, and produces results that have limited direct application. On the other hand, applied nutrition research tends to address immediate problems, to use human subjects, and to have limited control over the research setting but yields results that have direct value to practitioners. Christina (1989) suggested basic and applied research were useful in informing each other as to future research directions.

Field versus Laboratory Research

Field research is done outside the tightly controlled environment of the laboratory: a school, classroom, neighborhood community, park, hospital --- any location that exists in the "real world." Laboratory research is conducted under more "strict" conditions, which allows researchers to exert tighter control over an experiment. This control facilitates research being sound, but it can limit the application of the results. In real life, many factors may affect learning other than the experimental variable, and field research allows these other factors to operate. Some mistakenly feel that field research is inferior to laboratory research. This is certainly not true. Some settings make experimental research difficult if not impossible. For instance, it is unethical to introduce diseases into human subjects, so animal subjects are used. However, there are limitations to the ability to generalize the results of animal research to humans. This is certainly the case in many environments in which nutrition-related health professionals work. It would be more accurate to state that both experimental and field research have strengths as well as limitations. Field and laboratory research will be discussed in more detail in Chapters 11 and 12.

Qualitative Research, Descriptive Research, and Experimental Research

The term qualitative research is an umbrella term referring to several research traditions and strength that share certain commonalities. There is an emphasis on process, or how things happen, and focus on attitude, beliefs, and thoughts – how people make sense of their experiences as they interpret their world. The researcher does not manipulate variables through experimental treatment but takes more interest in process than in product. The researcher observes and gathers data in the field that is natural setting. Qualitative research has always been an integral part of cross-cultural comparisons and descriptions of food habits in the nutrition and anthropology literature.

Descriptive research is a study of status and widely used in education, nutrition, epidemiology, and the behavioral sciences. Its value is based on the premise that problems can be solved and practices improved through observations, analysis, and descriptions. The most common descriptive research method is the survey, e.g. nutrition survey. Developmental research is also descriptive. Through cross sectional and longitudinal studies, researchers investigate the interaction of diet and lifestyles and of disease development. The methods used are based mainly on comparative observations made at the level of whole populations, special groups (such as migrants, or vegetarians), or individuals within a population, who are investigated by methods using varying degrees of control.

Only experimental studies in which intervention is designed by the researchers provide proof of cause and effect. Experimental research is usually acknowledged as being the most scientific of all the types of research because the researcher can manipulate treatments to cause things to happen (i.e. a cause-and-effect situation can be established).

Unscientific and Scientific Methods of Problem Solving

Before we go into more detail concerning the scientific method of problem solving, it is important to recognize some other ways by which humankind has acquired knowledge. All of us have used these methods, so they are recognizable. Helmstadter (1970) labeled the methods as tenacity, intuition, authority, the rationalistic method, and empirical method.

Tenacity

People sometimes cling to certain beliefs despite the lack of supporting evidence. Our superstitions are good examples of this method called "tenacity." Take, for example, the man who believed that black cats brought bad luck. If one holds tenaciously to one's beliefs, even in the face of evidence that casts doubt on their validity, one seems to strengthen the beliefs.

Intuition

The a priori method is the way of knowing or fixing belief. Cohen and Nagel call it the method of intuition. Intuitive knowledge is sometimes considered to be common sense or self-evident. However, many self-evident truths are subsequently found to be false. That the earth is flat is a classic example of the intuitively obvious; and that the sun is farther away in winter than in summer was once self-evident.

Authority

This is the method of established belief. If the Bible says it, it is so. If a prominent professor says that modern education is soft and bad, it is so. Reference to some authority has long been used as a source of knowledge. Although this is not necessarily invalid, it does depend on the authority and on the rigidity of adherence. Even personal observation and experience have been deemed unacceptable when they dispute authority. Supposedly, people refused to look through Galileo's telescope when he disputed Ptolemy's explanation of the world and the heavens. Galileo was later jailed and forced to recant his beliefs.

Actually, life could not go on without the method of authority. We must take a large body of facts and information on the basis of authority. Thus, it should not be concluded that the method of authority is unsound; it is only unsound under certain circumstances.

The Rationalistic Method

In the rationalistic method, we derive knowledge through reasoning. A good example is the following classic syllogism:

All omega-3 fatty acids can reduce plasma triglycerides. (Major premise)
Docosahexaenoic acid is an omega-3 fatty acid. (Minor premise)
Therefore, docosahexaenoic acid reduces plasma triglycerides. (Conclusion)

Although you probably would not argue with this reasoning, the key to this method is the validity or veracity of the premise and their relationship to each other. For example,

Blacks have higher bone mass than Whites for comparable age and sex.
Tom is black man, and Mike is white man. Both of them are 20 years old.
Therefore, Tom has heavier bone mass than does Mike.

In this case, however, Tom has lower bone mass than Mike. The conclusion is trustworthy only if derived from premises (assumptions) that are true.

The Empirical Method

The word empirical denotes experience and the gathering of data. Certainly, data gathering is part of the scientific method of solving problems. However, there can be pitfalls in relying too much on your own experience (or data). First, your own experience (or data) is very limited. Furthermore, your retention depends substantially on how the events agree with your past experience and beliefs, on whether things "make sense," and on your state of motivation to remember. Nevertheless, the use of data (and the empirical method) is high on the continuum of methods of obtaining knowledge as long as you are aware of the limitations of relying too heavily on this method.

Empirical decisions are often based on pragmatic observations. Steroid and other performance-enhancing drug use by athletes is often the result of an empirical decision. These athletes may observe the total effect of steroid use in others without questioning its limitations. They do not solicit the deeper questions with regard to rapid strength and muscular gains steroid use provides. They are only interested in the end result. They tend to over-generalize.

The Scientific Method of Problem Solving

The methods of acquiring knowledge previously discussed lack the objectivity and control that characterize the scientific approach to problem solving. Several basic steps are involved in the scientific method. The steps are briefly described next.

Step 1: Developing the Problem (defining and delimiting it)

The researcher must be very specific about what is to be studied and to what extent it will be studied. Many ramifications constitute this step, an important one being the identification of the independent and the dependent variables. The **independent variable** is what the researcher is manipulating. If, for example, the effects of high and low fat diets on blood cholesterol levels are being compared, then the high and low fat diets are the independent variables; these are sometimes called the experimental, or treatment variables. The categorical variable is a kind of independent variable that cannot be manipulated because it is categorized by age, race, sex, and so on. If, for example, the effects of dietary fiber on blood cholesterol levels are being compared by age and sex, then age and sex are categorical variables.

The **dependent variable** is the effect of the independent variables. In the comparison of low and high fat diets, the blood cholesterol levels are the dependent variables. If you think of an experiment as a cause-and-effect proposition, the cause is the independent variable and the effect is the dependent variable. The latter is sometimes referred to as the yield. Thus, the researcher must define exactly what will be studied and what will be the measured effect. When this is resolved, the experimental design can be determined.

Step 2: Formulating the Hypothesis

The hypothesis is the expected result. When a person sets out to conduct a study, he or she generally has an idea as to what the outcome will be. This anticipated solution to the problem may be based on some theoretical construct, on the results of previous studies, or perhaps on the experimenter's past experience and observations. The research should have some experimental hypothesis about each sub-problem in the study.

Example 1-1:

Osteoporosis is one of the major health problems in postmenopausal women. The authors of the Framingham study have concluded that alcohol intake of at least 7 oz/wk is associated with high bone density in postmenopausal women, an effect possibly related to the augmentation of exogenous estrogen levels by alcohol (Felson 1995).

Problem: Is moderate alcohol consumption beneficial to postmenopausal women, especially, to osteoporotic women?

Hypothesis: Moderate alcohol intakes (slightly more than 7 oz/wk) increase bone density by increasing estrogen activity in postmenopausal women.

Step 3: Gathering the Data

The reliability of the measuring instruments, the controls that are employed, and the overall objectivity and precision of the data-gathering process are crucial to the problem solution.

In terms of difficulty, gathering data may be the easiest step because in many cases it is routine. However, planning the method is one of the most difficult steps. Good methods attempt to maximize both internal validity and external validity. In the above alcohol example, bone density and circulating estrogen levels should be measured by using the most sensitive instruments and authorized methodology.

Internal Validity --- The extent to which the results of a study can be attributed to the treatments used in the study.

External Validity --- The generalizability of the results of a study.

Step 4: Analyzing and Interpreting Results

The novice researcher finds this step to be the most formidable for several reasons. First, this step usually involves some statistical analysis, and the novice researcher has a limited background and a fear of statistics. Second, analysis and interpretation require considerable knowledge, experience, and insight, which the novice may lack.

The analysis and interpretation of results are the most challenging step without question. It is here that the researcher must provide evidence for the support or the rejection of the hypothesis. In doing this, the researcher also compares the results with those of others and perhaps attempts to relate and integrate the results into some theoretical model.

Example 1-2:
The Scientific Method

State the Problem
A. Does soy protein or milk protein have a greater impact on bone density?

1. State a testable or measurable hypothesis
- Soy protein has greater effects on prevention of osteoporosis than milk protein.

2. Plan the methods to be used in carrying out the study.
- Who will be the subjects?
- What will be their characteristics regarding age, sex, diet, physical activity, initial bone density, medical history, and so on?
- Which experimental design will be used? Cross-over design?
- How many groups will be in the experiment?

3. What measurements will be made? How and when will they be made?
- What are the best indicators for bone study?
  Bone density? Bone bio-markers?
- What is the duration of the experiment? etc.

4. Define exactly what the groups in the case will actually do?
- What will the diet consist of for each group?
- How will adherence to the diet be assessed?

5. How will the data be treated statistically?
- Student t-test?
- Analysis of variance?

6. Carry out the study.
- The two groups follow the guidelines for diet sets by the study.

7. Analyze the data using appropriate statistics.

8. State conclusions
- Soy protein significantly increased bone density more than did milk protein (This is just example of conclusion, not concluding from the author's experiment).

9. State a new research question.
- What are the mechanisms responsible for this conclusion?

Summary

Research is a way of solving problems. A good research project should be well planned, use up-to date technology, carefully analyze and accurately report the data, and appropriately deal with the ethical considerations associated with animals and humans when they are involved. There are basic and applied research; field versus laboratory research; and descriptive and experimental research. There are unscientific and scientific methods of problem solving. Several basic steps are involved in the scientific method: 1) developing the problem; 2) formulating the hypothesis; 3) gathering the data, and 4) analyzing and interpreting results.

We have presented here an overview of the nature of research. We identified different types of research. These categories and the different techniques that they encompass will be covered in detail in later chapters.

References

Austin, J.H. Chase, Chance and Creativity: The Lucky Art of Novelty. New York: Columbia University Press, 1978.

Bailey, K. Methods of Social Research. New York: Free Press, 1982.

Beveridge, W.B. The Art of Scientific Investigation. London: Heinemann, 1950.

Bickman, L. "Observational methods." In Research Methods in Social Relations. Selitz, C., Wrightsman, L., Cook, S. eds. New York: Holt, Rinehart, Winston, 1976.

Bogdan, R., Biklin, S. Qualitative Research for Education. Boston, MA: Allyn and Bacon, 1982.

Bogdan, R., Taylor, S. Introduction to Qualitative Research Methods. New York: Wiley, 1975.

Campbell, D.T. Stanley, J.C. Experimental and Quasi-experimental Design. Chicago, IL: Rand McNally, 1963.

Christina, R.W. Whatever Happened to Applied Research in Motor Learning? In Future Directions in Exercise and Sports Science Research. J.S. Skinner et al eds, Champaign, IL: Human Kinetics, 1989.

Cohen, M., Nagel E. An Introduction to Logic and Scientific Method. New York: Hartcourt, 1934.

Cook, T., Campbell, D. The Design and Conduct of Quasi-experiments in Field Settings. In Organizational Psychology. Dunnette, M. ed. Chicago, IL: Rand McNally, 1976.

Day, R.D. How to Write and Publish a Scientific Paper (2nd ed.) Philadelphia, PA: ISI Press, 1983.

Felson D.T., Zhang Y., Hannan M.T., Kannel W.B. Kiel D.P. Alcohol intake and bone mineral density in elderly men and women; The Framingham study. Am. J. Epidemiol 1995; 142:485-492

Helmstadter, G.C. Research Concepts in Human Behavior. New York: Appleton-Century-Crofts. 1970.

Hollman, A. Sir. Thomas Lewis: Pioneer Cardiologist and Clinical Scientist. London: Springer, 1996.

Kahn C.R. Sounding board. Picking a research problem: The critical decision. N Engl J Med, 1994; 330:1530-1553

Lequesne, M., Wilhelm, F. Methodology for Clinician. Basel: Eular, 1989.

Locke L.F. Qualitative research as a form of scientific inquiry in sport and physical education. Research Quarterly Exercise Sport, 1989; 60:1-20

Martens R. Science, knowledge, and sport psychology. Sports Psychologist, 1987; 1:29-55

Oyster, C.K., Hanten, W.P., Llorens, L.A. Introduction to Research: a Guide for the Health Science Professional. Philadelphia, PA: J.B. Lippincott, 1987.

Polanyi, M. Personal Knowledge. Chicago, IL: University of Chicago Press, 1958.

Weatherall, M. Medical research and national economics. J R Soc Med, 1981; 74: 407-408

Wilson, J. Thinking with Concepts. Cambridge, MA: Cambridge University Press, 1966.

CHAPTER 2

===

RESEARCH
PROBLEM AND LITERATURE REVIEW

Introduction

Getting started is the hardest part of almost any new venture, and research is no exception. You cannot do any significant research until you have identified the area you want to investigate, learned what has been published in that area, and figured out how you are going to conduct the investigation. In this chapter, we will discuss ways by which a person can identify researchable problems, search for literature, and write the literature review.

Identifying the Research Problem

Of the many major issues facing the graduate student, a primary one is the identification of a research problem. Problems may arise from real-world settings or be generated from theoretical frameworks. The source of research problems will vary according to the experience of the person contemplating an investigation, but it is generally agreed that the process begins with a question or need.

There are many reasons why people engage in research as delineated by Fox. Curiosity is as good a motivational factor as any. A graduate student was interested in the broad problem area of sweetener use in diabetic subjects. This student was curious about whether fructose could be used as a sweetener for diabetic subjects. She has learned that fructose ingestion causes less of an increase in blood glucose in normal and diabetic subjects than starch, sucrose or glucose. Fructose is metabolized mainly by the liver independent of insulin and no change in the insulin level is discernible after its use. By contrast, glucose metabolism is dependent on the presence of insulin secretion. On the other hand, she has also learned that fructose has been found to cause abnormal lipid patterns in humans, particularly increased serum triglycerides. She was so curious about pros and cons of fructose in diabetic subjects. Such curiosity led to an excellent study in which an attempt was made to determine the effect of long-term oral feeding of fructose on glucose and lipid metabolism in diabetic and control subjects (Ard). So from a problem area, a definitive problem was identified.

Before you begin to write a research proposal, some general as well as specific reading should be done. From this reading a general knowledge of the topic can be gained. This knowledge should include a familiarization with areas of controversy, designs, methods, and characteristics of the subject studied, as well as those not yet studied, and recommendations made by others. If the literature is reviewed with some of these aspects in mind, the results of your reading should be fruitful. Typically, graduate students tend to rush into reading the literature without much direction, which may require reading the some information.

Using Inductive and Deductive Reasoning

The means for identifying specific research problems comes from two methods of reasoning: **Deductive and Inductive.** Figure 2.1 provides a model of deductive reasoning. Deductive reasoning moves from a theoretical explanation of events to specific hypotheses that are tested against reality to evaluate whether the hypotheses are correct.

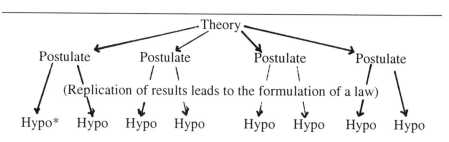

Comparison with reality

*Hypothesis
Figure 2.1: Deductive Reasoning
From a Student Guide for Educational Research (p23), R.L. Hoenes, B.S. Chissom, 1975, Statesboro, GA:Vog Press. Copyright 1975 by Vog Press.

Experimental-type (controlled setting, theory testing, predict) researchers use deductive reasoning and begin with the acceptance of a general principle or belief and then apply that principle to explain a specific case or phenomenon. This approach in research involves "drawing out" or verifying what already is accepted as true ((Hones & Chissom, 1975). For example, a researcher may start from the theory that stress increases blood pressures and that stroke problems emerge when such stressors as a lack of knowledge of controlling blood pressures.

Accepting these principles as true, the deductive researcher is interested in testing effectiveness of a series of interventions to reduce stress on blood pressures as a means of improving their health status.

Figure 2.2 provides a scheme of the inductive reasoning process. Individual observations are tied together into specific hypotheses, which are grouped into more general explanations that are united into theory. To move from the level of observations to that of theory requires many individual studies that test specific hypotheses.

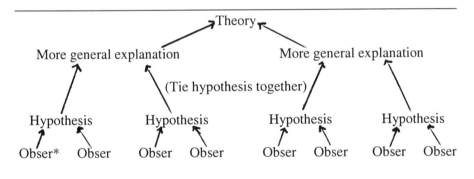

Figure 2.2: Inductive Reasoning
A Student Guide for Educational Research (p22) R.L. Hoenes, B.S. Chissom, 1975, Statesboro, GA: Vog Press. Copyright 1975 by Vog Press.

Researchers who work within a naturalistic framework (natural setting, reveal complexity, theory generating) primarily use inductive reasoning. This type of cognitive activity involves a process in which general rules evolve or develop from individual cases or observations of phenomena. Consider the same example of "stress and hypertension." The inductive researcher would be more interested in examining the relationship of stress intervening. From this approach, the researcher would develop a sense of what type of intervention would be most effective in reducing stress, consequently, blood pressures and stroke. The researcher, proceeding inductively, seeks to reveal or uncover a truth based on the observations of stress effects on blood pressures and stroke. Intervention principles then would be developed based on these observations.

In fact, within any given study, both inductive and deductive reasoning are useful. The researcher deduces hypotheses from relevant theories and concepts and induces hypotheses from relevant findings in other research.

Inductive Reasoning: an example

If 50 vegetarians were studied and all were found to be introverts, one might hypothesize that all vegetarians are introverts.

Deductive Reasoning: an example

All omega-3 fatty acids reduce blood triglycerides. Docosahexaenoic acid (DHA) is an omega-3 fatty acid. Therefore, DHA can reduce blood triglycerides.

Purposes of the Literature Review

A major part of developing the research problem is reading what has already been published about the problem. There may already have been much research done on the problem in which you are interested. In other words, the problem has been pretty much fished out. In many cases, your major professor can steer you away from a saturated topic. Whatever, the topic, past research is invaluable in planning new research. The purposes of literature review are: 1) to identify the problem; 2) to develop hypotheses; 3) to develop the method in your new research; and 4) to build on the body of knowledge.

Primary and Secondary References

References are either primary or secondary. Primary references are the original article, report, or book; secondary references are those in which the original work is described or mentioned by someone other than the author of the original work. In the latter case, one is informed secondhand. Much of the information stated in textbooks is based on original work that is described, cited in the text, and referenced. The text is thus a secondary reference for that particular information. The difference should be understood because, when writing a research paper or proposal, it is assumed that references cited were read firsthand to provide the best possible understanding of the original work. Restriction to secondary references increases the probability that opinions and bias are transferred from author to reader.

To ensure the highest level of accuracy in reporting information, the original or primary reference should be read firsthand. As a result, the references cited by you in a paper indicate that you did read the original work rather than a translation or summary of the work by another author. Consequently, secondary references should be used only when the original work is not available.

However, secondary sources such as textbook and encyclopedias are helpful when students have very limited knowledge about a topic and will profit from background information and a summary of previous research. A review paper on the topic of interest is especially valuable.

Encyclopedias provide an overview of information on research topics and summarize knowledge about subject areas. Review of research, i.e. **Annual Review of Nutrition, Nutrition Review,** etc. are an excellent source of information.

PRELIMINARY SOURCES: Abstracts, Indexes, Bibliographies, and Internet Sources

Use preliminary sources to find primary sources via computer-aided hand searches. Preliminary sources primarily consist of abstracts and indexes.

Abstracts: Concise summaries of research studies are valuable sources of information. Abstracts of papers published, and of papers presented at research meetings are available at international, national, district, and state conventions. Medline has abstracts written by authors, whereas other indexes have independent abstracts.

Indexes: The Medicus Index provides access to more than 2,500 biomedical journals around the world. It is published monthly, and each issue has subject and author sections and a bibliography of medical reviews. Index Medicus can also be searched via computer through Medline.

Bibliographies: Bibliographies list books and articles about specific topics. They come in many forms, depending on how the information is listed. All contain the authors, titles of books or articles, journal names, and publishing information. A good search strategy is to look for the most recent sources as information and then work backward.

The catalog is a type of bibliography but it often provides links to full text documents, graphics and other primary sources.

The Library Information System

Most university libraries have gone to a computerized catalog system. Usually, the searcher first selects the type of search from a menu, such as author, title, key word, or call number. Many libraries in colleges and universities can also be accessed by faculty and students by personal computer via modem.

Computer Searches

Computer service facilities can greatly expedite the literature search. Automated searching provides more effective and efficient access to indexes and information than does manual searching.

The Internet provides continually improving access to a massive and expending pool of information including international nutritional biomedical and clinical sources. Not only does it link to "traditional" bibliographic information services as mentioned above, but it offers through the World Wide Web (WWW) a varied medium for the retrieval of audio-visual media and textual and graphical documentation from a range of organizations including universities, governments, research institutes and hospitals.

The web pages are the most current and easiest to access form of information. For instances, the web page sites: http://www.diabetes.org/nutrition/ for the American Diabetes Association; http://www.faseb.org/asns for the American Society for Nutritional Sciences; http://www.faseb.org/ascn for the American Society for Clinical Nutrition; and http://www.eatright.org for the American Dietetic Association.

WWW browsing software packages like Netscape have "hotlinks" or bookmark facilities so that you can build your own library of locations of value.

Electronic Journals

The electronic journals offer a network presence that may consist of table of contents listing abstracts, job advertisements, general news and instruction to authors, the key biomedical databases. While some may offer full text access, and will ultimately require a subscription payment to access the full service. The Journal of Nutrition is an example.

Finally, you may consult a librarian for additional relevant databases on technical search skills to improve retrieval reference with specificity.

Obtaining the Primary Sources

After you have a list of related references, you must obtain the actual studies and read them. Many references have abstracts in addition to the bibliographic information. The abstract is extremely helpful in making the decision as to whether the article is worth retrieving.

Keep in mind that a computer search does not replace the hand search. The computer is remarkably fast and effective in identifying references that may be pertinent to one's topic. However, perhaps the most valuable step in the search process is finding a recent, closely related study and reading that study's review of literature. Then, you find those studies that were cited and you read them, which leads you to other sources.

Your library may not have all journals that are on the list of references. You must consult your library information system to see if your library carries particular journals. If it doesn't you will want to use interlibrary loan or possibly other document delivery options such as the FAX of the article. Electronic mail has fostered a rapid and simple means of getting the article.

Reading and Recording the Literature

Collecting related literature is a major undertaking, but the next step is even more time-consuming. You must read, understand, and record the relevant information from the literature. When studies are particularly relevant to the proposed research, make a photocopy. Write the complete citation on the title page if the journal does not provide this. Many students make the mistake of simply photocopying an article without looking to see if the pages copied contain all the necessary information for a citation.

The primary elements of a bibliographic reference usually include the name of the author, the title, the place of publication, the publisher's name, the date of publication, and page number of a reference.

To summarize, the best system for recording relevant literature is probably a combination of note taking and photocopying. By using index cards (4"x 6" cards are usually large enough), the important information about most studies can be recorded and indexed by topic. Always be sure to record the complete and correct citation on the card with the appropriate citation style. Or all correct information can be inputted in the computer for later citations.

There are citation manager software packages for downloading a literature search and keeping these in a particular style. The most popular one is:

Literature Review Chart

Author (date)	Sample Boundaries	Design	Independent Variables	Dependent Variables	Results
Lee et al (1989)	10 wk old rats Male Sprague-Dawley	2x2 ANOVA	Phytoestrogen Alcohol	Bone Density	Increase Increase

Citation: Journal Article: Journal, Author, Title, Vol Number, Page Numbers, & Date
Book: Chapter author & title, Book author & title, Edition, Page number, Date & Publisher

Personal bibliographic packages like reference manager, Papyrus, Procite and Endnote are becoming essential research tools. They enable you to store, retrieve and edit references, prepare bibliographies and effortlessly incorporate them into the word-processed body of your papers. Citations downloaded from electronic databases like Medline and EMBASE can be imported into your own personal database and then manipulated accordingly.

Critically Evaluating the Literature

There are several guiding questions you can use to help you critically evaluate the literature. Use the questions in Table 2.1 to guide your reading of research literature. Your responses to these evaluative questions will inform your research direction.

Table 2.1: Questions for Analysis of Research

Overall Impression
1. Is the paper a significant contribution to knowledge about area?
2. Is the study clear, and internally consistent?

Introduction
1. Is the research plan developed within a reasonable theoretical framework?
2. Is current and relevant research cited and properly interpreted?
3. Is the statement of the problem clear, concise, testable, and derived from the theory and research reviewed?

Method
1. Are relevant subject characteristics described, and are the subjects appropriate for the research?
2. Is the instrumentation appropriate?
3. Are testing or treatment procedures described in sufficient detail?
4. Has been the protocol of the study approved?
5. Are the statistical analyses and research design sufficient?

Results
1. Do the results evaluate the stated problem?
2. Is the presentation of results complete?
3. Are the tables and figures appropriate?

Discussion
1. Are the results discussed?
2. Are the results related back to the problem, theory, and previous findings?
3. Are the conclusions supported by the study?

References
1. Are all references in the correct format, and are they complete?
2. Are all references cited in the text?
3. Are all dates in the references correct, and do they match the text citation?

Abstract
1. Does it include a statement of the purpose; description of subjects, instrumentation, and procedures; and a report of meaningful findings?
2. Is the abstract the proper length?

General
1. Are key words provided?
2. Are running heads provided?
3. Does the paper provide for use of nonsexist language, protection of human subjects or animals, and appropriate labeling of human subjects?

Thomas, J.R. & Nelson, J.K. Research Methods in Physical Activity (p41), used with permission.

Writing the Literature Review

After the notes have been taken from all of the relevant and pertinent sources, they should be sorted and classified. The researcher should gain insight into agreements, differences, relationships, and trends as a result of this process. When two or more literary sources do not agree, both sides of the issue should be noted. Beginning researchers sometimes have the mistaken idea that they should seek out only those hypotheses, findings, and conclusions with which they agree.

The literature review has three basic parts: introduction; body; and summary and conclusions. The introduction should explain the purpose of the review and the how and why of its organization. The body of the literature review requires considerable attention. Relevant research must be organized, synthesized, written in a clear, concise, and interesting way. The literature review should be organized around important topics. These topics serve as subheadings in the paper to direct the reader's attention.

Notice that the review is organized around concepts of themes, rather than the one study per paragraph approach. A review of osteoporosis, for example, may be organized into subheadings on bone structures, bone turnover, effects of estrogen on bone mass, effects of testosterone on bone mass, effects of vitamin D on bone mass, effects of calcium on bone mass, etc. Several paragraphs under each topic will depict the overall findings and cite individual studies to document the observations. For example, a sentence in one paragraph may be stated as follows: "Several studies have indicated that parathyroid hormone increases with aging."

A more interesting and readable approach is to present a concept and then discuss the various findings about that concept, documenting findings by references to the various research reports related to it. In this way, consensus and controversy can be identified and discussed in the literature review. More relevant and important studies can be presented in greater detail, and several studies with the same outcome can be covered in one sentence.

No one can just sit down and write a good literature review. A careful plan is necessary. The best way to organize the topics and the information within topics is to develop an outline. The more carefully the outline is planned, the easier the writing will be. A good task is to select a review paper from a journal or from a thesis or dissertation review of literature and reconstruct the outline the author must have used.

Suggested Outline for Writing a Review

   I. Introduction (overview of what the review covers)
  II. Review of specific concepts
      A. How each concept has been studied
         1. Overview of studies
           a. Design
           b. Results
           c. Critical evaluation
         2. Critical evaluation of current knowledge
 III. Integration of concepts
      A. Relationships proposed in studies
      B. Identification of conflicts in the literature
      C. Identification of gaps in the literature
      D. Identification of research needs
 IV. Rationale for study and design
  V. Overview

Summary

Identifying and formulating researchable problem is difficult task for the graduate student. Some suggestions were given to help the graduate student find suitable topics. Inductive and deductive reasoning were discussed with regard to formulating research hypotheses. A major part of developing the research problem is reading what has already been published about the problem. The purposes of literature review are 1) to identify the problem; 2) to develop hypotheses; 3) to develop the method in your new research; and 4) to build on the body of knowledge. Relevant research must be organized, synthesized, written in a clear, concise, and interesting way. The literature review should be organized around important topics.

There are no shortcuts to locating, reading, and indexing the literature and then writing the literature review. To write a good literature review, a careful plan is necessary. The best way to organize the topics and the information within topics is to develop an outline. The more carefully the outline is planned, the easier the writing will be.

References

Ard, N.F. Long term effects of oral fructose on blood lipid components in diabetic and non-diabetic subjects. Thesis, Norman, Oklahoma; University of Oklahoma, 1984.

Davidson, M. Computing and Information Technology. London: Straightforward Publishing, Ltd, 1994.

Day, R.D. How to Write and Publish a Scientific Paper. (2nd ed) Philadelphia, PA: ISI press, 1983.

Glowniak J.V., Bushway M.K. Computer networks as a medical resource: Accessing and using the internet. JAMA, 1994; 271: 1934-1939.

Fox, D.J. The Research Process in Education. New York: Holt, Rinehart & Winston, 1969.

High Wire Press: Internet imprint of the Stanford University Libraries. http://www.highwire.org/

Hones, R.L., Chissom, B.S. A Student Guide for Educational Research. Statesboro, GA: Vog Press, 1975.

Koh E.T., Ard N.F., Mendoza F. Effect of fructose feeding on blood parameters and blood pressure in impaired glucose-tolerant subjects. J Am Dietet A, 1988; 88:932

Kroll, E. The Whole Internet: User's Guide and Catalog. Sebastopol, O'Reilly and Associates, 1994.

Lee, R. How to Find Information – Life Science: a Guide to Searching in Published Source. London: British Library, 1992.

Loehle C. A guide to increased creativity in research - inspiration or perspiration? BioScience, 1990; 40:123 -129

Lowe H.J., Barnett G.O. Understanding and using the medical subject headings (MeSH) vocabulary to perform literature searches. JAMA, 1994; 271: 1103-1108

Morse, J. Qualitative Nursing Research. Rockville, MD: Aspen, 1989.

McKenzie, B. Medicine and the Internet: Introducing Online Resources (4th ed.) London: Bowker Saur, 1992.

Snyder, C.W. Jr, Abernethy, B. eds. The Creative Side of Experimentation. Champaign, IL: Human Kinetics, 1992.

Thomas, J.R., Nelson, J.K. Research Methods in Physical Activity (3rd ed.) Champaign, IL: Human Kinetics, 1996.

Van Dalen, D.B. Understanding Educational Research: An Introduction (4th ed.) New York: McGraw-Hill, 1979.

Woodsmall R.M., Benson D.A. Information resources at the National Center for Biotechnology Information. Bulletin Medical Library Association, 1993; 81:282-284

CHAPTER 3

===

FRAMING A RESEARCH PROBLEM: HYPOTHESES, PURPOSES, OBJECTIVES, AND QUESTIONS

The problem chapter provides an overview of the study: statement of the purpose or problem to be addressed, title, introduction, hypothesis, delimitations (scope), limitations (variables that could not be controlled), and significance of the study.

Choosing the Title

The purpose of the title is to convey the content of the study, but this should be done as succinctly as possible. In another word, the cause- and effect-relationship of the independent and dependent variables should be described in the title.

Although discussing the title first may seem logical, it might surprise you to learn that titles are often not determined until after the study has been written. However, at the proposal meeting, you must have a title (even though it may be provisional), so you will discuss it first. Usually, the title in the proposal can be changed when you are writing your thesis.

The following are principles for writing titles for empirical research reports (Pyrczak & Bruce, 1992).

1. If only a small number of variables are studied, the title should name the variables.
2. If many variables are studied, only the type of variables should be named.
3. A title should indicate what was studied with or without the results of the study.

This principle may surprise some beginning students of empirical methods because outcomes and conclusions are often stated in titles in the popular journals. (See the following examples)

4. Mention the population(s) in a title when the type(s) of population(s) are important.
5. Consider the use of subtitles to amplify the purposes or methods of study.
6. A title may be stated in the form of a question: this form should be used sparingly and with caution.
7. A title should be consistent with the research hypothesis, purpose, or question.

Here are some examples in nutrition areas:

From the Journal of the American Dietetic Association:

- Primary follow-up care in a multi- disciplinary setting enhances catch-up growth of very-low-birth-weight infants

- Profitability and acceptability of fat- and sodium-modified hot entrees in a work-site cafeteria

- Beneficial lipid outcome of medical nutrition therapy for men with combined hyperlipidemia in an ambulatory setting

From the Journal of Nutrition:

- Protein synthesis and degradation change rapidly in response to food intake in muscle of food-deprived mice

- Chronic exercise affects vitamin B-6 metabolism but not requirement of growing rats

- Dietary skim milk powder increases ionized calcium in the small intestine of piglets compared to dietary defatted soybean flour

You will recognize outcomes and conclusions are stated in most of titles in the Journal of Nutrition, but not in the Journal of American Dietetic Association.

From the Journal of Epidemiology

- Assessment of trans-fatty acid intake with a food frequency questionnaire and validation with adipose tissue levels of trans-fatty acids

- Smoking, lipids, glucose intolerance, and blood pressure as risk factors for peripheral arteriosclerosis compared with ischemic heart disease in the Edingurg artery study

From Theses of Graduate Students (University of Oklahoma):

- Comparison of nutrient intakes of postmenopausal women in relation to race and risk factors for heart disease (1998)

- The effects of a high and low intensity weight training program on bone mineral density in early postmenopausal women (1998)

- The effects of fluoride, albumin, and a combination of fluoride and albumin on mineral deposition in bone and selected soft tissues of non-mated and mated female mice and their progeny (1997)

You will notice that the titles in the journals consist of a complete sentence (mainly in the Journal of Nutrition), while those in the theses do not. However, there is no restriction on this; the title does not end with a period mark, appropriate to titles.

Writing the Introduction

The purpose of an introduction in an **empirical research** report is to introduce the problem area, establish its significance, and indicate the author's perspectives on the problem. In other words, introductions usually conclude with an explicit statement of the research hypotheses, purposes, or questions to be answered by the study.

In a journal article, the introduction is almost always integrated with the literature review into a single essay. Most institutions of higher education require that the introduction and the review of literature in a thesis or dissertation be presented in separate chapters.

A good introduction requires literary skill because it should flow smoothly yet be reasonably brief. The introductory paragraphs must create interest in the study: thus your writing skill and knowledge of the topic are especially valuable in the introduction. The narrative should introduce the necessary background information quickly and explain the rationale behind the study. A smooth, unified well-written introduction should lead to the problem statement with such clarity that the reader could state the study's purpose before specifically reading it.

The following examples specify some desirable features in an introduction, including a general introduction, background information, a mention of gaps in the literature and areas of needed research, and logical progression leading to the problem statement.

Example 3-1:

A chick bioassay approach for determining the bioavailable choline concentration in normal and overheated soybean meal, canola meal and peanut meal (Emmert & Baker, J Nutr 127:745-752, 1997) Used with permission.

General introduction

Choline is essential for the prevention of fatty liver and perosis in poultry, and it has long been known that chicks fed corn-soybean meal diets containing (total) choline in excess of the NRC (1994) requirement still need supplemental choline to achieve maximum growth.

Background introduction

The obvious implication is that choline bioavailability is <100% in these diets. Previous estimates of choline bioavailability in soybean meal have ranged from 60% to 75%, and choline in canola meal, although present at high concentrations, has also been found to be less than fully available. However, attempts to quantify choline bioavailability have met with criticism, primarily because of the use of weight gain as the response criterion for chicks fed choline-deficient diets. In addition, weight gains of chicks fed purified crystalline amino acid diets, such as have been used previously, may respond to components other than choline in ingredients such as soybean meal (SBM).

Lead-in

Therefore, it is imperative that an appropriate basal diet be used that will allow reliable estimates of choline bioavailability from various feed ingredients.

We have previously developed an experimental diet containing soy protein isolate (SPI) and 2-amino-2-methyl-1-propanol that is singularly deficient in choline and will elicit a weight gain response only upon addition of choline. {Our objective in this study was to estimate the bioavailable concentration and percentage bioavailability of choline in soybean meal, canola meal and peanut meal. In addition, the effects of overheating on choline bioavalability were assessed.}

Example 3-2:

Maternal intake of cruciferous vegetables and other foods and colic symptoms in exclusively breast-fed infants (Lust et al., J Amer Dietet Assoc, 96: 46-48, 1996) Used with permission.

General introduction

There is nothing in the field of pediatrics that is more bothersome to the physician, aggravating to parents and more painful to infants than persistent colic in new borns.

Through the work of Spock and Aldrich in the 1940s and Illingworth and Wessel in the 1950s, the clinical definition of colic emerged. Colic occurs when a healthy infant develops recurrent paroxysms of abdominal pain beginning by the second or third week after birth; it usually abates by the fourth month of life. Colic is characterized by incessant, inconsolable crying lasting 3 hrs or more per day and occurring for more than 3 days in any one week. It is generally accompanied by the symptoms of irritability, gastrointestinal distention, and abdominal cramping. Colic tends to occur during the late afternoon or early evening. Estimates of the prevalence of infantile colic range from 10% to 40% of US infants aged 0 to 4 months

Background introduction

A number of etiologic theories have been proposed for infantile colic, but causative factors have yet to be determined. Suspected causes of infantile colic include poor parent-infant interactions and foods such as cow's milk and soy protein.

Lead-in

To our knowledge, a relationship between maternal intake of cruciferous vegetables and colic in exclusively breast-fed infant has not been reported in the scientific literature, although cruciferous vegetables are apparently regarded as being highly suspicious in causing colic.

{Our study investigated the relationship between consumption of cruciferous vegetables (e.g., cabbage, Brussels sprouts, cauliflower, and broccoli) and other foods by exclusively breast-feeding women and the prevalence of one or more of the major symptoms of colic infants. We hypothesized that colic symptoms would be associated with maternal intake of cruciferous vegetables.}

Example 3-3

General Introduction

In recent years, there has been growing interest in eating disorders in athletes. Studies have shown that athletes are more prone to developing eating disorders than non-athletes. In addition, the highest prevalence of eating disorders is in female athletes competing in sports where leanness and/or a specific weight are considered important for either performance or appearance.

Background Information

There has been considerable speculation about why athletes are at increased risk for eating disorders. Predisposing personality or family interaction variables might be primary, so participation in sports favoring leanness could be a consequence of preexisting eating problems or could be coincidental. Alternatively, participation in certain sports could be related causally to the onset of eating disorders. In all likelihood these factors interact.

Lead-In

One important area of inquiry, therefore, is to identify risk factors to help determine which athletes are most vulnerable, or which conditions or trigger factors elicit the pathological behavior.

Problem Statement (Purpose of the study)

To examine risk factors for eating disorders along with trigger factors that may be responsible for precipitating the onset or exacerbation of eating disorders (in elite female athletes).

Framing a Research Problem

An endless number of problems or issues can emerge from any one topic. The challenge to the researcher is to identify one particular area of concern or specific research problem within a broad topic area. A specific research problem is a statement that identifies the phenomenon to be explored and why it needs to be examined or why it is a problem or issue. You may feel that because there are so many problems to be explored that it is very simple to select one for research. However, how one frames and states the problem is critical to the entire research endeavor and influences all subsequent thought and action processes. That is, the way in which a problem is framed determines the way in which it will be answered. Therefore, this first research step should be thought through very carefully.

To move from the selection of a broad topic area to framing a concrete research problem, the following questions can help guide your thinking:

1. What about this topic is of interest to me?
2. What about this topic is of relevance to my practice?
3. What about this topic is controversial and unsolved in the literature?
4. What about this topic is of importance or significance in social or scientific aspects?

By answering these reflective questions, you will begin to narrow your focus and hone in on a researchable problem.

Stating the Research Problem

The problem statement follows the introduction. We should point out that the literature review is often included in the introductory section and thus precedes the formal statement of the problem. If this is the case, then a brief problem statement should appear fairly soon in the introductory section before the literature review.

The problem statement in the above example 3-1 was to estimate the bioavailable concentration and percentage bioavailability of choline in soybean meal, canola meal and peanut meal; and to assess the effects of overheating on choline bioavailability.

The problem statement in the previous example 3-2 was to assess relationships between consumption of cruciferous vegetables and other foods by exclusively breast-feeding women and the prevalence of colic symptoms in infants.

Identifying the Variables

The statement should identify the different variables in the study, including the independent variable, the dependent variable, and the categorical variable (if any). Usually, some control variables (which could possibly influence the results and are kept out of the study) can also be identified.

The independent and dependent variables have already been mentioned in Chapter 1. The independent variable is the experimental, or treatment variable: it is the "cause". The dependent variable is what is measured to assess the effects of the independent variable; it is the "effect". A categorical variable is sometimes called a moderator variable. This variable is a kind of independent variable, except that it cannot be manipulated because it is categorized by, for example, age, race, and sex. It is studied to determine whether the cause-and-effect relationship of the independent and dependent variables is different in the presence of the categorical variable or variables.

Example 3-3:

Cholesterol-lowering effects of modified animal fats in postmenopausal women:

Independent variable --- Modified animal fats
Dependent variable ----- Cholesterol
Categorical variable ---- Postmenopausal women

Example 3-4:

Dietary lipids modulate bone prostaglandin E2 production, insulin-like growth factor-1 concentration and formation rate in chicks.

Independent variable --- Dietary lipids
Dependent variables ---- Prostaglandin production, insulin-like growth factor
                          concentration and formation rate
Categorical variable ---- Chicks

The researcher decides which variables to manipulate and which variables to control. One can control the possible influence of some variable by keeping it out of the study. Thus, the researcher chooses not to assess a variable's possible effect on the relationship between the independent and dependent variables, so this variable is controlled. For example, suppose a researcher is comparing dietary treatments on cholesterol levels in the elderly. The researchers have a choice. They can include men and women for subjects as a categorical variable, or they can control the variable of sex by including men only.

The decision to include or exclude some variable depends on several considerations, such as whether the variable is closely related to the theoretical model and how likely there is to be an interaction. Practical considerations include how difficult it is to make a variable a categorical variable or to control it and how much control the researcher has over the experimental situation.

Quantitative and Qualitative Variables

Variables can be qualitative if they are classified by some characteristic, attribute, or property. People are categorized as to sex (male, female), eye color (blue, brown, green), church preference (Catholic, Protestant, Jew), and political affiliation (Republican, Democrat, Independent). Qualitative variables are usually unmeasurable. Variables can also be quantified or measured in a numerical sense. For example, heights of children can be measured in centimeters. There are two basic types of quantitative variables, discrete and continuous.

1. Discrete Variables --- This type of variable is usually thought of as being a whole unit, one that cannot be fractionated or divided up into smaller parts. Examples of discrete variables are football scores, and the number of correct answers on a test. Football scores are recorded in whole numbers and cannot be divided into smaller parts like 7.5 or 9.8.

2. Continuous Variables --- This type of variable can be divided into fractional amounts in large or small degrees. Strength and endurance scores, height, weight, and girth measures are considered to be continuous variables. My height is 165.5 cm.

Hypotheses

After you have stated the research problem, you must present the hypothesis. A hypothesis is a statement of the likely outcome of a study. It is often posed as a question. It can also be thought of as a prediction of findings based on the relevant literature.

Statistical Hypotheses: Null and Directional

Statistical hypotheses are stated in either a null form or a directional form. In the former no significant effect or relationship is anticipated. For example, using the example of body fatness and TV watching, the null hypothesis might be stated as "Body fatness and time spent watching TV will not be significantly related." The directional hypothesis is stated such that a significant difference or relationship is predicted to occur. In this case, TV watching would be projected to be significantly related to body fatness.

Research Hypotheses

Research hypotheses are stated according to the results the researcher actually expects. They might appear as follows: The high fat diet increases more cholesterol than does the low fat diet. Research hypotheses are used in this section of the research paper. Statistical hypotheses are used to test the research hypothesis. Hypotheses are not usually stated in journal article because they are implied by the purpose of the study. Using the fat diet experiment as an example, the purpose of the study would be to compare the effect of high- and low-fat diets on plasma cholesterol levels. Logically the research hypothesis then states what the expected outcome will be. For the above example, "High-fat diets increase more plasma cholesterol as compared to low-fat diets."

Operationally Defining Your Terms

Another task in the preparation of the first section of a thesis or dissertation is operationally defining certain terms so that the researcher and the reader can adequately evaluate the results. It is imperative that the dependent variable is operationally defined.

Two types of definitions are usually found in empirical research reports. Conceptual definitions, which refer to the general concepts, are often presented in the introduction. Operational definitions, which define traits in concrete, step-by-step physical terms, are usually presented in the section on methods.

In theses and dissertations, conceptual definitions are often presented in a separate section of the introduction, with its own subheading. In journal articles, conceptual definitions are usually integrated into the introductory statement. Often, authors of journal articles assume that their readers are familiar with the general concepts and, thus, do not provide formal statements of conceptual definitions. In both types of reports, operational definitions should be provided in the section of methods.

Outlining Basic Assumptions, Delimitations, and Limitations

Besides writing the introduction, stating the research problem and hypothesis, and operationally defining your terms, you must outline the basic assumptions and limitations under which you performed your research.

Assumptions

Every study has certain fundamental premises without which it could not be conducted. An assumption is believed to be fact, but cannot be verified as one.

In the case, for example, of nutrition surveys using questionnaires, the researcher may make the assumption that the subjects answer the questionnaire honestly.

Limitations

A limitation is either (1) a weakness or handicap that potentially limits the validity of the results or (2) a boundary to which the study was knowingly confined. The latter is often called a delimitation. Limitations are uncontrollable events that may interfere with the results of a study. One common limitation is the duration of the study. Short-term studies may be limited simply because some phenomena do not dramatically change in 10- to 12-weeks. Studies examining bone density change through exercise or diet should last many months because bone responds slowly to these stimuli. A 12-week study may conclude falsely that these variables have no effect on bone simply because of the inappropriately short duration of the study.

Stating limitations in a distinct section of the proposal ensures that the student recognizes the possible influence of other variables. No study can perfectly control all factors in the environment, and therefore even the best designed and best planned studies have one or more limitations. In published articles, limitations are often mentioned in the discussion section or are implied.

Delimitations

For example, if a researcher wanted to study nutritional status of children, but used only a measure of caloric intake, this would be a limitation in the first sense because it is a weakness in the execution of the study. If the researcher wanted only to study caloric intake as part of nutritional status, his findings would be delimited to caloric intake, which is not a flaw in light of the investigator's purpose.

Delimitations are the "what, who, where of a study and serve to summarize what is included in the study: the nature of the subjects, the location of the study, its duration, and variables studied." An example might be: Eighty-five 7- and 8-year-old children enrolled at Kennedy Elementary School in Oklahoma City will be subjects in a 12-week weight reduction study. In journal articles this information is covered in the methods section of the paper.

Explicit statements of assumptions, limitations, and delimitations are usually required in theses and dissertations. These statements are usually included in Chapter 1, each with its own subheading. However, these are not mandatory requirements for theses or dissertations.

Authors of journal articles often integrate these statements in various sections of their articles, including the introduction, method section, and discussion section. They often are very selective in deciding which assumptions and limitations to state, naming only the major ones.

Summary

This chapter discussed an overview of the study: statement of the purpose or problem to be addressed, title, introduction, hypotheses, delimitations, limitations, and significance of the study.

The purpose of the title is to convey the content, but this should be done as succinctly as possible. The purpose of an introduction in an empirical research report is to introduce the problem area, establish its significance, and indicate the author's perspectives on the problem. A good introduction requires literary skill because it should flow smoothly yet be reasonably brief. The introductory paragraphs must create interest in the study: thus your writing skill and knowledge of the topic are especially valuable in the introduction. The challenge to the researcher is to identify one particular area of concern or specific research problem within a broad topic area. The problem statement follows the introduction. The statement should identify the different variables in the study, including the independent, the dependent, and the categorical variables.

Another task in the preparation of the first section of a thesis or dissertation is operationally defining certain terms so that the researcher and the reader can adequately evaluate the results. There are two types of definitions: conceptual and operational definitions. Besides writing the introduction, stating problem and hypothesis, and operationally defining your terms, you must outline the basic assumptions and limitations under which you performed your research.

40

References

DePoy, E., Gitlin, L.N. Introduction to Research: Multiple Strategies for Health and Human Services. St Louis, MO: Mosby, 1993.

Emmert J.L., Baker D.H. Chick bioassay approach for determining the bioavailable choline concentration in normal and overheated soybean meal, canola meal and peanut meal. J Nutr 1997; 127:745-752

Friedman, L.M., Furberg, C.D., DeMets, D.L. Fundamentals of Clinical Trials. (2nd ed.) Littleton, MA: PSG Publishing Co, Inc., 1985.

Herbert P.C., Tugwell P.K. A reader's guide to the medical literature – an introduction. Postgrad Med J 1996; 72: 1-5

Hulley, S.B., Cummings, S.R. Designing Clinical Research: An Epidemiologic Approach. Baltimore, MD: Williams & Wilkins, 1988.

Kuhn, T.S. The Structure of Scientific Revolutions. Chicago, IL: University of Chicago Press, 1962.

Lequesne, M., Wilhelm, F. Methodology for the Clinician. Basel: Eular, 1989.

Lust K.D., Brown J.E., Thomas W. Maternal intake of cruciferous vegetables and other foods and colic symptoms in exclusively breast-fed infants. J Am Dietet A 1996; 96::46-48

Polit, D., Hungler, B. Nursing Research: Principles and Methods, (2nd ed.) Philadelphia, PA: J.B. Lippincott Co., 1983.

Pyrczak, F., Bruce, R.R. Writing Empirical Research Reports: A Basic Guide for Students of the Social and Behavioral Sciences. Los Angeles, CA: Pyrczak Publishing, 1992.

Silverman, W.A. Human Experimentation: A Guided Step into the Unknown. Oxford, England: Oxford University Press, 1985.

Tuckman, B.W. Conducting Educational Research (2nd ed.) New York: Harcourt Brace Jovanovich, 1978.

CHAPTER 4

======================================

WRITING METHOD SECTIONS

Planning the Work

The question to be answered or the hypothesis to be tested now has to be turned into a practical project, which will consist of making observations on some form of biomedical material, performing clinical trials on patients, or carrying out experiments on animals or in vitro. The planning stage for the research should be described in the method section.

This chapter will describe an overview of the research method. The methods section has one major goal. If someone else follows exactly what is described in the methodology section, they should get the identical results, and confirm the findings of the original researcher. If they do not, then the methods were not precise enough or there was a methodological error. Therefore, it is better to over describe than under describe the methods employed, especially in the first draft of a report.

When reading published articles, many readers may skip the methods section, jumping to the results to see what happened. However, it is very important to read the methods section first. This allows the reader to understand fully what was attempted and allows a better appreciation of the results. In addition, the methods section of other articles may provide the researcher with information on a more appropriate methodology for his or her study.

The methods section contains a description of the physical steps taken to gather the data. Typically, it begins with a description of the **subjects, instrumentations and apparatus,** (i.e., measuring tools). Any additional **procedures such as administration of experimental treatments, design and analysis** should also be described here. In reporting on completed research, use the past tense to describe methods; in proposals, use the future tense.

In the methods, you must determine:

1. the type of subjects, with criteria for inclusion and exclusion
2. the source of the subjects and how they are selected
3. the number of subjects and why that number was chosen
4. the type of observation and the measurement techniques to be used

Describing Subjects

This section of the method of a thesis or dissertation describes how and why the subjects were selected and which of their characteristics are pertinent to the study. The subjects should be described in enough detail so that the reader can visualize the subjects. Issues to consider when selecting subjects revolve around these questions:

Specification: The type of subjects with criteria for inclusion and exclusion

Suppose an investigator studies the efficacy of calcium supplementation with magnesium, boron, silicon and vitamin D for prevention of osteoporosis in postmenopausal women. He or she has to create two sets of selection criteria, inclusion and exclusion criteria at the beginning of the study, which will define the study populations. These criteria should be described in the method section.

Inclusion Criteria: The inclusion criteria define the main characteristics of the target and accessible populations. The task of specifying the clinical characteristics involves difficult judgments about what factors are important to the research questions.

Example 4-1 Inclusion Criteria

For the above osteoporosis study, the investigator should define "post-menopausal women" by clinical characteristics and ages. For instance, the inclusion criteria include women at least one year from the last menopausal period or women with bilateral ovariectomy or, for women who have had a simple hysterectomy as age of at least 54 years. Age range will be between 45 and 60 years, and all subjects must be ambulatory.

Exclusion Criteria: Exclusion criteria indicate subsets of individuals who meet the eligibity criteria, but are likely to interfere with the quality of the data or the interpretation of the findings.

Example 4-2 Exclusion Criteria

Any clinical conditions and any drugs taken which may affect bone mass and metabolism should be considered in planning the osteoporosis study. Thus, the exclusion criteria include for the bone study: 1) currently treated major disease of organ systems including liver, kidneys, heart, or lungs; 2) current neurologic, hematologic, or bone disease; 3) current or past history of all cancers; 4) current major psychiatric or endocrine disorders (e.g. thyroid, parathyroid, etc.); current medications that might interfere with bone metabolism (e.g. anticonvulsants glucocorticoids, etc.) and all anti-osteoporotic drugs (e.g. calcitonin, ipriflavone, sodium fluoride, etc.); and current use of raloxifen, or tamoxifen.

For example, the patient with liver or kidney disease should be excluded because the two organs involve vitamin D activation and metabolism, which directly influence bone metabolism. Another example, thyroid and parathyroid glands directly involve bone formation and resorption. Therefore, all these clinical conditions may interfere with the quality of bone data or the interpretation of the findings on the efficacy of calcium supplementation for prevention of osteoporosis

The inclusion/exclusion criteria are more important in applied research than basic research. These criteria should be described in detail in the method section.

Additional special characteristics necessary for human subjects description in your research are:

1. ethnic and cultural backgrounds
2. age (children, elderly)
3. gender (male, female)
4. weight
5. health status (other clinical characteristics than those described in inclusion/exclusion criteria)
6. educational levels (e.g.,12th grade, nutritional knowledge)
7. income levels (poverty income, middle income)
8. trained or untrained (level of training)
9. experts or novices (level of performances)

10. size (weight, fatness)
11. exact number of subjects, attrition, etc.

Subjects Sources and Selections

Where and how subjects will be identified or selected should be described in the method section. The sources may be a particular school, retirement communities, shopping mall, or hospitals, etc. For example, you compare the incidence of bone fractures among different ethnic postmenopausal women such as American Indians, Asians, and Whites. You may identify American Indian subjects at the Indian Public Health Services, Indian Hospitals, and Indian tribal offices. These are excellent sources of sampling for American Indians. You may identify Asian subjects at the Asian Society or Asian communities.

Now, you should describe the recruitment strategy and sampling methodology. Usually the accessible population is too large, and there is a need to select a small group of individuals for study. A population is an all-inclusive group that is operationally defined by the researcher. A sample is a representative subset of the population that contains the essential elements of that population. Anytime a researcher is interested in making inferences he or she needs to have a sample that is a good representation of the population. How does one obtain a good model sample? There are two main classes of sampling designs, probability and nonprobability. A probability sample uses a random process to guarantee that each unit of the population has a specified chance of selection. Nonprobability sampling designs are more practical than probability design for many clinical research projects. We will discuss about sampling more in detail in Chapters 6 and 7.

An important factor to consider in choosing the accessible population and sampling approach is the feasibility of recruiting the subjects into the study. There are two main goals: (1) to recruit enough subjects to meet the sample size requirements of the study, and, (2) to recruit a sample that is unbiased.

In clinical research it is sometimes possible to avoid sampling and its biases by studying the entire accessible population – all cases of a rare disease. This is the best approach when it is feasible.

The sampling procedures and sample size should be described in detail, along with discussion of errors that compromise the validity of applying the study's conclusions to the target populations.

Who to Tell About the Subjects

The exact number of subjects should be given, as should any loss of subjects during the time of the study. If there was attrition, state the number of subjects who dropped out, the reasons for the attrition, if known, and information about the drop-outs, if available. In the proposal, some of this information may not be exact.

Example 4-3: Detailed description of the subjects

Title: Effects of Fructose Feeding on Blood Parameters and Blood Pressure in Impaired Glucose-Tolerant Subjects (Koh et al., J Amer Dietet Assoc, 88:932-938, 1988)

Nine impaired glucose-tolerant subjects (IGTS), three men and six women, were selected using the following criteria: 1- to 2- hour postprandial plasma glucose levels between 160 and 200 mg/dL (8.8 to 11.1 mmol/L); compliance with the diet; and no complications from renal and heart diseases. All subjects were outpatients at a private clinic and had well adhered to their dietary prescriptions before the study. All IGTS had been treated by diet only, under the direction of the dietitians who conducted the present study.

The control subjects (NGTS) were selected from members of the university faculty and staff, who understood the nature of the study and the importance of complying with the diet. They were matched by race, sex, and age (±5 years) with the selected patients. Also, the control subjects were not allowed to use medicine during the study period. (Table 4.1)

The researchers in this study were fortunate that there was no attrition in the sample size.

Table 4.1: Comparison of anthropometric measurements and ages of IGTS and NGTS

Variables	NGTS(n=9)	IGTS(n=9)
Weight (lb)	145 ±10*	164 ±11
Height (in)	66 ±1	65 ± 1
Subscapular skinfolds (mm)	14.2±1.2	22.7 ± 4.9
Triceps skinfolds (mm)	13.7±1.7	19.5 ± 3.1
Arm circumference (cm)	27.6±0.8	31.8 ± 1.3
Pulse (beats/min)	70 ±2	74 ± 3
Age (yrs)	50 ±5	54 ± 6

*Mean±SD

Protecting Subjects

Most research in the study of physical activity, nutrient and drug treatments deals with humans, often children, elderly, or patients, but it also includes animals. Therefore, the researcher must be concerned about any circumstances in the research setting or treatment that could harm the humans or animals. In the next Chapter, "Ethical Issues in Research and Scholarship," we provide details on what the researcher must do to protect both humans and animals used in research. It is particularly important to obtain informed consent of humans and ensure the protection and care of animals.

Of course, informed consent is not always required. For example, if you are conducting an observational study of the developmental behaviors of children in a nursery school, you may not need to obtain informed consent for that observation.

For such particular exception, the subject protection description like "The experimental protocol was approved by the Human Subject Committee, Institutional Review Board, at the University of America" should be included in the methods.

Animal Study:

For an animal experimental study you must determine:

1. selection of animals or tissue
2. selection of controls
3. selection of experimental methods

When using animal subjects in the study, the same principles should be applied as those of human subjects. The following information should be provided.

Animals:
1. source – the suppliers of experimental animals
2. species
3. breed or strains
4. age
5. weight
6. special conditions, etc.

If animals are used as subjects for research studies in life science, institutions require adherence to the Guide for the Care and Use of Laboratory Animals, published by the U.S. Department of Health and Human Services, as detailed in the Animal Welfare Act (PL 89-544, PL 91-979, and PL 04-279).

Example 4-4: Animal Subjects

Title: Low Dietary Protein Impairs Blood Coagulation in BHE/cdb Rats. (Chang et al., J Nutr 127:1279-1283, 1997)

Animal Care and Diet Composition

Specific pathogen-free, male BHE/cdb rats from the colony at the University of Georgia (Berbanier 1995) were used according to procedures for care and experimental protocol approved by the University of Georgia Institutional Animal Care and Use Committee. The rats were 2-3 months old at the beginning of the experiments, and were housed individually in wire-bottom cages in a room controlled for temperature ($21\pm1^{\circ}$), humidity (40-50%), and light (lights on, 0600-1800). Diets were isocaloric, with sucrose substituted to compensate for differences in protein (casein : lactalbumin, 1 : 1 by weight), and fat was contributed by 1g corn oil/100g diet and either 9g beef tallow or 9g menhaden oil/100g diet, as indicated (Table 4.2).

Diet and Other Interventions

It is best to present the diet composition or other interventions in tabular form. This allows for easy comparison of treatments, and encourages the readers to think of the relevant components of the intervention. The proximate composition of closed formula diets should be given as amounts of protein, energy, fat and fiber. Components should be expressed in mass concentrations (g/kg diet) or substance concentrations (µmol/kg diet). It is necessary to state the method of delivery of the diet (Table 4.2). The mean daily intake by each group should be present unless pair-fed.

Example 4-6: From the same article (Chang et al.,1997)

Table 4.2: Diet Composition

	Protein		Fat			Carbohydrate Sources
	Lactalbumin	Casein	Corn Oil	BT	MO	
			g/kg			
80g/kg P+BT	40.0	40.0	10.0	90.0	---	725.0
80g/kg P+MO	40.0	40.0	10.0	---	90.0	725.0
380g/kgP+BT	190.0	190.0	10.0	90.0	---	425.0
380g/kgP+MO	190.0	190.0	10.0	---	90.0	425.0
120g/kgP+BT	60.0	60.0	10.0	90.0	---	685.0
180g/kgP+BT	90.0	90.0	10.0	90.0	---	625.0
240g/kgP+BT	120.0	120.0	10.0	90.0	---	565.0
300g/kgP+BT	150.0	150.0	10.0	90.0	---	505.0

1All diets contained per kg: 10g AIN-76 vitamin mixture; 3.50g AIN-76 mineral mixture; 50g Celufil and 0.00040g DL-α-tocopheryl acetate

2P, protein content (w/w); BT, beef tallow; and MO, menhaden oil

Describing Instruments

Information about the instruments, apparatus, or tests used to collect data is used to generate the dependent variables in the study. Consider the following points when selecting tests and instruments:

1. What is the validity and reliability of the measures (see Chapter 10)?
2. How difficult is it to obtain the measures?
3. Do you have access to the instruments, tests, or apparatus needed?
4. Do you know (or can you learn) how to administer the tests or use the equipment?
5. Do you know how to evaluate subjects' test performance?
6. Will the tests instruments, or apparatus yield a reasonable range of scores for the subjects you have selected?
7. Will the subjects be willing to spend whatever time is required for you to administer the tests or instruments?
8. What are the models, and manufactures' names and locations of mechanical instruments?

Your major professor may suggest what instruments to use for your study.

Describing Procedures

In this section you should describe how the data are obtained, including all testing procedures for obtaining scores on the variables of interest. How tests are given and who gives them are important features. You should detail the step of the testing situation and instructions given to the subjects (although you may place some of this information in the appendix). If the study is experimental, then you should describe the treatments applied to the different groups of subjects. Consider these points when planning procedures:

Collecting the data:

1. When? Where? How much time is required?
2. Do you have pilot data to demonstrate your skill and knowledge in using the tests and the equipment and how subjects will respond?
3. Have you developed a scheme for data acquisition, recording, and scoring?

Planning the treatments:

1. How long? How intense? How often?
2. How will subjects' adherence to treatments be determined?
3. Do you have pilot data to show how subjects will respond to the treatments and that you can administer these treatments?
4. Have you selected the appropriate treatments for the type of subjects to be used?

Describing Design and Analysis

Design is the key to controlling the outcomes from experimental and quasi-experimental research. The independent variables are manipulated in an attempt to judge their effects on the dependent variable. A well-designed study is one in which the only explanation for change in the dependent variable is how the subjects were treated (independent variable). The design and theory have enabled the researcher to eliminate all rival or alternative hypotheses using the following MAXICON principles (Thomas & Nelson, 1996). The design requires a section heading in the method for experimental and quasi-experimental research. The experimental designs will be discussed in Chapter 11.

MAXICON Principles

1. Maximize true variance, or increase the odds that the real relationship or explanation will be discovered.
2. Minimize error variance, or reduce all the mistakes that could creep into the study to disguise the true relationship.
3. Control extraneous variance, or make sure that rival hypotheses are not the real explanations of relationship.

Statistical Analysis

The plans for data analysis must also be reported. In most studies some type of statistical analysis is used, but there are exceptions (e.g. historical or qualitative research). Therefore, researchers should provide the following information on statistical analyses at the end of the methods and material section:

1. data presentation, i.e. means and standard deviations (SD) or standard error of the mean (SEM) for each variable
2. analytical procedures, i.e., *t*-test, correlation and regression analysis, analysis of variance, etc.
3. whether variables were transformed
4. significant test, i.e., Duncan's multiple test, Newman-Keul's test, Tukey's test, etc.
5. significant levels, i.e., $p < 0.01$ or $p < 0.05$, etc.
6. program used for analyses, i.e., SAS, SPSS, etc.

For example, if correlational techniques (relationships among variables) are used, then the variables to be correlated and the techniques are named: "The degree of relationship between two estimates of percent fat will be established by using Pearson *r* to correlate the sum of three skinfolds with underwater weighing" ($p < 0.05$). (More in the following examples)

Examine the following Methods and Materials for the four elements described: subjects, instruments or apparatus, procedures, and design and analysis. Also state all operational definitions in the methods and materials. What are the assumptions, limitations, and delimitations?

Example 4-7: Animal study

Title: Comparison of copper status in rats when dietary fructose is replaced by either cornstarch or glucose (Koh, P.S.E.B.M., 194:108-113, 1990)

Animals

Weanling male Sprague-Dawley rats (Hilltop Lab Animals, Scottsdale, PA) weighing approximately 50-60 g each were housed individually in stainless steel cages with were mesh bottom in a temperature humidity controlled room with a 12-hr light/dark cycle. The rats were randomly divided into 10 groups. They were fed either sole fructose as the carbohydrate source (100% fructose); 50% starch and 50% fructose (50% starch); 50% glucose and 50% fructose (50% glucose); 75% starch and 25% fructose (75% starch), 75% glucose and 25% fructose (75% glucose) with or without copper.

Table 4.3: Experimental Groups

Carbohydrate in diet			Copper	Sample	Group
Fructose (%)	Starch (%)	Glucose (%)	(μg/g)	No.	Designation
62.8			7.06	8	+Cu100F*
62.8			0.79	13	- Cu 100F
31.4	31.4		7.52	8	+Cu 50S
31.4	31.4		1.01	8	- Cu 50S
15.7	47.1		7.07	8	+Cu 75S
15.7	47.1		0.98	8	- Cu 75S
31.4		31.4	7.42	8	+Cu 50G
31.4		31.4	0.96	8	- Cu 50G
15.7		47.1	7.81	8	+Cu 75G
15.7		47.1	0.89	8	- Cu 75G

Copper level analyzed by atomic absorption

+Cu100F =One-hundred percent fructose with copper

- Cu100F =One-hundred percent fructose without copper

+Cu 50S = Fifty percent fructose replaced by starch with copper

- Cu 50S = Fifty percent fructose replaced by starch without copper

+Cu75S = Seventy-five percent fructose replaced by starch with copper

- Cu75S = Seventy-five percent fructose replaced by starch without copper

+Cu 50G = Fifty percent fructose replaced by glucose with copper

- Cu 50G = Fifty percent fructose replaced by glucose without copper, so on

Diets

The basal diet contained the following (g/kg): 628 carbohydrate, 200 eggwhite; 95 corn oil, 30 cellulose; 35 AIN mineral mix prepared in my laboratory and formulated to omit cupric compounds from the mineral mix for the copper-free basal diet, 2.7 choline bitartrate, and 10 vitamin AIN-76 supplemented with 2 mg of biotin/kg diet. The basal diets contained 0.79-0.98 ug of copper/g diet. Cupric carbonate (10.5 mg of CuCO3/kg) supplemented diets contained 7.06-7.81 ug of copper/g diet. All animals were provided with distilled, deionized drinking water. Body weight and food intake were measured weekly.

Procedures

The termination of the experiment was based on the time of death of the first rat. This occurred during the fifth week in the rats fed 100% fructose-copper diet. None of the animals in the other groups died. The study was terminated following an overnight fast. Animals were killed by decapitation and blood was collected into heparinized test tubes. The blood was centrifuged at 2200g for 20 min at 4°C and the plasma was analyzed for triglycerides, cholesterol, and ceruloplasmin. Erythrocytes were used to determine copper-zinc-superoxide dismutase activity. Livers and hearts were removed from the rats immediately after they were killed and weighed, washed in ice-cold saline solution, and homogenized with five volumes of 0.2% Triton X-100 using a homogenizer equipped with stainless steel blades. Homogenate aliquots were extracted to isolate copper-zinc-superoxide dismutase.

Chemical Assays

Plasma total cholesterol and triglycerides were determined by enzymatic methods using the Centrifichem System (Baker Instrument Co., Allentown, PA). Plasma ceruloplasmin was measured by the use of O-dianisidine dihydrochloride. Copper-zinc-superoxide dismutase activity was determined according to the method of Misra and Fridovich. Copper concentration in liver and heart was measured following their digestion by a method combining wet and dry ashing. The ashed residue was dissolved in 0.1 N HCl. Duplicate samples of the tissue homogenates and of plasma were analyzed for copper by flame atomic absorption spectrophotometry (model 5000; Perkin-Elmer, Norwalk, CT). Bovine liver 1577a from the National Bureau of Standards Reference Materials was digested and analyzed along with samples to verify accuracy. The percentage of recoveries for copper was 98.5%.

Statistics

Data were analyzed by analysis of variance (ANOVA) using the SAS software system for data analysis. In all of the statistical comparisons, difference with $p < 0.05$ were considered to be significant.

Summary

This chapter has provided an overview of the method for research study. We have identified the major parts as subjects, instruments or apparatuses, procedures, and design and analysis. The four parts of the method chapter and their major purposes are to eliminate alternative or rival hypotheses or to control any explanation for the results except the hypothesis that researcher intends to evaluate. The MAXICON principle shows the way to accomplish this: a) maximize the true or planned sources of variations, b) minimize any error or unexplained sources of variation, and c) control any extraneous sources of variation. In the following sections, we will study how to do this from the viewpoints of statistics and of design.

References

Armitage, P. "Exclusions, Losses to Follow-up, and Withdrawals in Clinical Trials." In Clinical Trials: Issues and Approaches., Shapiro, S.H., Louis, T.A. eds., New York: Marcel Dekker, Inc., 1983.

Babbie, E. The Practice of Social Research. Belmont, CA: Wadsworth, 1983.

Bains C.J. Impediments to recruitment in the Canadian National Breast Screening Study: response and resolution. Cotr Clin Trials 1984; 5: 129-140

Bradford, H.A., Hill, I.D. A Short Textbook of Medical Statistics. (12th ed.) London: Hodder and Stoughton, 1991.

Chang Y-L, Shon H-S, Chan K-C, Berdanier C.D., Hargrove J.L. Low dietary protein impairs blood coagulation in BHE/cdb rats. J Nutr 1997; 127:1279-1283

Chalmers, I., Altman, D.G. eds. Systemic Reviews. London: British Medical Journal, 1995.

Cohen J. Things I have learned. Amer Psychologist 1990; 45:1304-1312

Ellenberg J.H., Nelson K.B. Sample selection and the natural history of disease. Studies of febrile seizures. JAMA 1980; 243:1337-1343

Gardner, M.J., Altman, D.G. Statistics with Confidence. London: British Medical Journal, 1989.

Kahn, H.A. An Introduction to Epidemiologic Methods. New York: Oxford University Press, 1983.

Kelsey, J.F., Thompson, W.D. Evans, A.S. Methods in Observational Epidemiology. New York: Oxford University Press, 1986.

Kelinbaum, D.G. Kupper, L.L., Morgenstern, H. Epidemiologic Research: Principles and Quantitative Methods. Belmont, CA: Lifetime Learning Publications, 1982.

Koh E.T., Ard N.F., Mendoza F. Effects of fructose feeding on blood parameters and blood pressure in impaired glucose-tolerant subjects. J Am Dietet Assoc 1988; 88: 932-938

56

Koh E.T. Comparison of copper status in rats when dietary fructose is replaced by either cornstarch or glucose. Proc Soc Exp Biol Med 1990; 194:108-113

Lipid Research Clinics Program: Recruitment for clinical trials. Circulation 1982; 66: (Suppl IV) 1-78

Marks, R.G. Designing a Research Project. Belmont, CA: Lifetime Learning Publications, 1982.

Melton L.J. Selection bias in the referral of patients and the natural history of surgical conditions. Mayo Clin Proc 1985; 60:880-889

Pocock, S.J. Clinical Trials: A Practical Approach. Chester, New York: John Wiley & Sons, 1983.

Polit, D. Hungler, B. Nursing Research: Principles and Methods, (2nd ed.) Philadelphia, PA: JB Lippincott Co., 1983.

Schlesselman, J.J. Case-Control Studies: Design, Conduct, Analysis. New York: Oxford University Press, 1982.

Spilker, B. Guide to Clinical Studies and Developing Protocols. New York: Raven Press, 1984.

Siemiatycki J., Campbell S., Richardson L., Aubert D. Quality of response in different population groups in mail and telephone surveys. Am J Epidemiol 1984; 120: 302-314

Thomas, J.R., Nelson, J.K. Research Methods in Physical Activity. (3rd ed.) Champaign, IL: Human Kinetics, 1996.

Whimster, W.F. Biomedical Research: How to Plan, Publish and Present It. London: Springer-Verlag, 1997.

CHAPTER 5

===

ETHICAL ISSUES
IN RESEARCH AND SCHOLARSHIP

Introduction

As graduate students, you will encounter a number of ethical issues in research and scholarship. In this chapter we draw your attention to many of these issues and provide a framework for discussion and decision making. However, the choices will not always be clear-cut. The most important aspect of making good decisions is to have good information and obtain advice from trusted faculty. The major topics to be presented include misconduct in science, working with faculty and other graduate students, using humans as subjects in research, and using animal subjects.

Misconduct in Science

Scientific research can offer the exhilaration of discovery to scientists. However, research can also entail frustrations and disappointments to scientists. An experiment may fail because of poor design, technical complications, or the sheer intractability of nature. A favored hypothesis may turn out to be incorrect after consuming months of effort. Colleagues may disagree over the validity of experimental data, the interpretation of results, or credit for work done. Difficulties such as these are virtually impossible to avoid in science.

Errors arising from human fallibility also occur in science. Even the most responsible scientist can make an honest mistake. When such errors are discovered, they should be acknowledged, preferably in the same journal in which the mistaken information was published. Scientists who make such acknowledgments promptly and openly are rarely condemned by colleagues.

Mistakes made through negligent work are treated more harshly. Some researchers may feel that the pressures on them are an inducement to haste at the expense of care. They may believe that they have to do substandard work to compile a long list of publications. Sacrificing quality to such pressures can easily backfire. Scientists with a reputation for publishing a work of dubious quality will generally find that all of their publications are viewed with skepticism by their colleagues.

Because of these erroneous reports, many other researchers can waste months or years of effort, and public confidence in integrity of science can be seriously undermined.

Beyond honest errors and errors caused negligence is a third category of errors: those that involve deception. They are various types of misconduct in science.

Misconduct includes fabrication (making up data or results), falsification (changing or misreporting data or results), plagiarism (using the ideas or words of another person without giving appropriate credit), or other practices that seriously deviated from those that are commonly accepted within the scientific community for proposing, conducting, or reporting research. Misconduct is unethical and subject to heavy penalties.

Within the scientific community, the effects of misconduct-in terms of lost time, forfeited recognition to others, and feelings of personal betrayal-can be devastating. Individuals, institutions, and even entire research fields can suffer grievous setbacks from instances of fabrication, falsification, or plagiarism even if they are only tangentially with the case

Plagiarism

Scientific publication is an important part of the process by which credit and priority are established for experimental work and research ideas. Duplicating without citation of texts previously published by others or expropriating without attribution the experimental findings, methods, or ideas of others is plagiarism and is unethical. Of course, this is completely unacceptable in the research process. Plagiarism carries severe penalties at all institutions. A researcher who plagiarizes work carries a stigma for life in his or her profession. Everyone involved in the publishing process, including authors, reviewers, and editors, has a responsibility to maintain maximum levels of honesty.

In scientific writing, originality is also important. Common practice is to circulate preprints and drafts of papers among scholars (which are often shared with graduate students) who are known to be working in a specific area. If ideas, methods, findings, and so on are borrowed from these, proper credit should always be given.

Fabrication and Falsification

There are approximately 40,000 journals that publish more than one million papers annually (Henderson, 1990). Thus, it is not surprising to learn that scientists have occasionally been caught making up or altering research data. Pressures have been particularly intense in medical and health-related research because such research is often expensive, requires outside funding, and involves risk. It seems easy to make a change here or there or to make up data.

Fabrication in a publication

On rare occasion graduate students and faculty may knowingly produce fraudulent research; established scholars are sometimes indirectly involved in scientific misconduct. This may occur in working with other scientists who produced fraudulent data that followed the predicted outcomes (e.g., as in a funded grant where the proposal had suggested what outcomes were probable). In these instances, the established scholar sees exactly what he or she expects to see in the data. Because this verifies the hypotheses, the data are assumed to be acceptable. For example, a case of established scholar John involved a paper published in the refereed journal in April 1986 signed by the scholar John and coauthors Bill, Steve and David as principal authors. The established scholar John had checked the findings of David, but he saw in the data the expected outcomes and agreed to submit the paper. The fact that the data were made up subsequently led to his resignation as president of R University. Thus, even though he was not the paper's principal author, his character was seriously damaged by being an unwilling party to scientific fraud.

Sometimes major professors expect certain outcomes of graduate students' experiments. However, when graduate students finished their experiments, the outcomes were opposite. Although the professor may be upset with the outcome, students should be honest with their data. Here is a graduate student's experience: Menoheptulose has been shown to inhibit the in vitro glucose-stimulated release of insulin by slices of pancreatic tissue. Similarly, the administration of menoheptulose via an intramuscular injection completely suppressed the increase in serum immunoreactive insulin in response to an oral glucose load. Therefore, the major professor expected decreased serum insulin in menoheptulose-injected rats as compared to the saline-injected control rats. However, the outcome was opposite. Of course, her major professor was upset with such unexpected data, and tried to repeat the experiment using another graduate student. However, the graduate student proved her honesty with extensive reference works, and published an excellent paper with those results in the Journal of Nutrition. Her major professor was happy at the end. Honesty is an essential element in scientific research.

Falsification can also occur with related literature. Graduate students should be careful in how they interpret what an author says. Work of other authors should not be bent to fit projected hypotheses. This is also a reason graduate students should read original sources instead of relying on the interpretations of others (secondary references), as those interpretations may not follow the original source closely.

Misleading Authorship

Definition of Authorship

According to the "Uniform Requirements" for Manuscripts Submitted to Biomedical Journals (N Engl J Med 324:424-428, 1991), the following are the principles by which to judge claims to authorship:

All persons designated as authors should qualify for authorship. The order of authorship should be a joint decision of the co-authors. Each author should have participated sufficiently in the work to take public responsibility for the content.

Authorship credit should be based only on substantial contributions to
1. conception and design, or analysis and interpretation of data
2. drafting the article or revising it critically for important intellectual content
3. final approval of the version to be published

Conditions 1,2, and 3 must all be met. Participation solely in the acquisition of funding or the collection of data does not justify authorship. General supervision of the research group is also not sufficient for authorship. Any part of an article critical to its main conclusion must be the responsibility of at least one author.

In the standard scientific paper, credit is explicitly acknowledged in three places: in the list of authors, in the acknowledgment of contributions from others, and in the list of references or citations.

The Order of Authorship

A major ethical issue among researchers involves joint research projects or, more specifically, the publication and presentation of joint research efforts. Generally, the order of authorship for presentations and publications should be based on the researchers' intellectual contributions to the project. The first, or senior, author is usually the researcher who developed the idea and plan for the research. Second and third authors are normally listed in the order of their contributions (Fine & Kurdek, 1993). However, conventions differ greatly among disciplines and among research groups. Sometimes the scientist with the greatest name recognition is listed first, whereas in other fields the research leader's name is always last. In some disciplines supervisors' names rarely appear on papers, while in others the professor's name appears on almost every paper that comes out of the lab. Some research groups and journals avoid these decisions by simply listing authors alphabetically.

Co-authorship

The co-authors of a paper should be all those persons who have made significant scientific contributions to the work reported and who share responsibility and accountability for the results. Other contributions should be indicated in a footnote or an "Acknowledgements" section. For instance, technicians do not necessarily become joint authors. Graduate students sometimes feel that, because they collect the data, they should be coauthors. Only when graduate students contribute to the planning, analysis, and write-up of the research report are they entitled to be listed as coauthors. This rule may not apply to graduate students who are paid on their work on a funded project. A good major professor involves his or her graduate students in all aspects of his or her research program; thus, these students frequently may serve as both coauthors and technicians.

Gratuitous or honorary authorship is neither appropriate nor ethically acceptable. Authorship is not a gift, but a right founded on substantial contribution to the resulting manuscript. Honorary authorship is a misrepresentation, implying an intellectual contribution that was not made.

Honorary authors risk associating themselves with work that may later be the subject of a misconduct investigation.

For instance, on June 7, 1996 (p1498), <u>Science</u> published a report entitled "Synergistic activation of estrogen receptor with combinations of environmental chemicals."

In the paper at issue, the Tulane team used yeast cells with a human estrogen receptor to test the potential estrogenic effects of different compounds. They found that pairs of several pesticides were 1000 times more potent at triggering estrogenic activity than were individual chemicals on their own. The prospect that pesticides could mimic the female sex hormone raised alarm bells among toxicologists and environmentalists and helped convince Congress to include provisions in two 1996 laws requiring manufactures to screen thousands of chemicals on the market for estrogenic activity. Within a few months after the Science article appeared, however, other labs reported that they could not replicate its results.

On July 18, 1997, the paper's senior author, John A. McLachlan, sent <u>Science</u> a letter formally withdrawing the report.

Concurrent with the withdrawal of the paper, Tulane University convened a committee of academicians to investigate possible scientific misconduct by Steven F. Arnold, the study's principal investigator, and McLachlan, the senior author. McLachlan is the director of the Center for Bioenvironmental Research at Tulane University. Arnold resigned from the university in 1997.

After a comprehensive review, the committee determined that McLachlan did not commit, participate in, or have any knowledge of any scientific misconduct. However, the fact that without any intellectual contribution he accepted senior authorship is unethical.

With respect to Arnold, the committee concluded that he provided insufficient data to support the major conclusions of the Science paper. Additionally, independent review of Arnold's data does not support the major conclusions contained within the <u>Science</u> paper. (Science 284 (5422): 1905 & 1932)

Ethical Issues Regarding Copyright

Graduate students should be aware of copyright regulations and the concept of "fair use" as it applies to published materials. Copyrighted material is often used in theses or dissertations, and this is acceptable if the use is fair and reasonable. Often graduate students will want to use a figure or table from another source. You must seek permission from the copyright holder (for a published paper usually the author, but sometimes the research journal) and cite it appropriately (e.g., used or reproduced with permission).

Protecting Human Subjects

Current guidelines and regulations regarding human experimentation have evolved over the last half century in reaction to public and scientific outcry over a few individual cases of gross human injustice. One of the first to receive public scrutiny was the heinous behavior of physicians toward the inmates of Nazi concentration campus in Germany during World War II. Following the war, 20 doctors were tried in Nuremberg before an international tribunal for war crimes and crimes against humanity. The resulting Nuremberg code of 1947 established ten principles that must be followed in human experimentation to satisfy moral, ethical and legal concepts. These principles, for the first time, established as essential informed voluntary consent of human subjects.

Established ten principles by Nuremberg Code are the following:

1. The voluntary consent of the human subject is absolutely essential.
2. The experiment should be such as to yield fruitful results for the good of society, unprocurable by other means or methods of study and not random and unnecessary in nature.
3. The experiment should be so designed and based on the results of animal experimentation and a knowledge of the natural history of the disease or other problem under study that the anticipated results with justify the performance of the experiment.
4. The experiment should be so conducted as to avoid all unnecessary physical and mental suffering and injury.
5. No experiment should be conducted where there is an a priori reason to believe that death or disabling injury will occur; except, perhaps, in those experiments where the experimental physicians also serve as subjects.
6. The degree of risk to be taken should never exceed that determined by the humanitarian importance of the problem to be solved by the experiment.
7. Proper preparations should be made and adequate facilities provided to protect the experimental subject against even remote possibilities of injury.

8. The experiment should be conducted only by scientifically qualified persons. The highest degree of skill and care should be required through all stages of the experiment of those who conduct or engage in the experiment.
9. During the course of the experiment the human subject should be at liberty to bring the experiment to an end if he has reached the physical or mental state where continuation of the experiment seems to him to be impossible.
10. During the course of the experiment the scientist in charge must be prepared to terminate the experiment at any stage, if he has probable cause to believe, in the exercise of the good faith, superior skill, and careful judgment required of him that a continuation of the experiment is likely to result in injury, disability or death to the experimental subject.

The second major international code of ethics was the Declaration of Helsinki adopted by the World Medical Association in 1964, with a proviso that the text be reviewed periodically. The basic principles in the Declaration of Helsinki were extended by the 29th World Medical Assembly in Tokyo in 1975, and further revised in 1983. The 12 basic principles delineated the concept of submitting experimental protocols to an independent committee for consideration comment, and guidance. This was the genesis of the future institutional review board (**IRB**), which has become a major force in ensuring the right of human subjects. The Helsinki Declaration also counseled researchers to exercise caution in conducting research that could affect the environment and to respect the welfare of animals used for research.

A third important document supporting the rights of human subjects is the 1978 Belmont Report issued by the National Commission for the Protection of Human Subjects of Biomedical and Behavioral Research (National Commission). This President's Commission was formed in 1974 in rapid response to disclosure of two scandalous research studies of the 1930s. One a study of the immune response, involved the injection of live, malignant cells into several aged patients in a chronic disease hospital without the patients' prior consent. The second study concerned long term observation of the "natural" course of syphilis in 300 Black man who were recruited into the study without informed consent. More despicably, these men were observed for several decades but did not receive penicillin, the efficacy of which in the treatment of syphilis was established several years after the initiation of the study.

The Belmont Report addresses ethical conduct of research involving human subjects and argues for balancing society's interests in protecting the rights of subjects with its interests in furthering knowledge that can benefit society as a whole. To assess benefit and risk of any research protocol, three basic principles are evoked: respect for persons, beneficence, and justice.

The reports issued by the National Commission established recommendations for the protection of special categories of human subjects, including the human fetus, children, prisoners, and people institutionalized as mentally infirm. To protect the rights of subjects, Institutional Review Board (**IRB**) were empowered through federal regulations.

In all aspects of human experimentations, it is critical that researchers avoid misrepresentation to human subjects. The paramount strategy to avoid this research error relies on the three ethical principles of respect, beneficence, and justice. These principles affirm full and comprehensible disclosure to subjects, noncoercive consent, autonomous right of free choice, including the right to terminate participation without penalty, confidentiality, protection of privacy, and equitable selection of subjects. A research project should be terminated if, at any point, the data warrant such action.

All of the federal guidelines and regulations promulgated since 1971 have mandated that an institutionally sponsored and locally based committee accept responsibility for protecting the rights and safety of human subjects of biomedical research. Each revision of and addition to the regulations has increased the amount and kind of responsibility thrust upon **IRB** members. Initially, the task of **IRB** members was threefold. First, the board was required to assure that the rights and welfare of the subjects were adequately protected. The second task was to assure that the risks to the individual were outweighed by the potential benefits to him, or by the importance of the knowledge to be gained. The third duty of the **IRB** was to assure that the informed consent was obtained by adequate and appropriate methods. The basic elements of informed consent are the following:

1. A fair explanation of the procedures to be followed, including an identification of
   the experimental elements.
2. A description of the attendant discomforts and risk involved.
3. A description of the benefits to be expressed.
4. A disclosure of appropriate alternative procedures that would be disadvantageous to the subject.
5. An offer to answer any inquires concerning the procedures.
6. An instruction that the subject was free to withdraw his consent and to discontinue participation in the project or activity at any time.

The new regulations impose greater responsibilities upon the IRB members. Among these is the task of ascertaining the acceptability of research proposals in terms of institutional commitments and regulations, applicable law and regulations, as well as standards of professional conduct and practice. Additionally the IRB must examine and judge the appropriateness of the proposed research design and the merit of the study in question. Moreover, the new regulations invest the IRB with the explicit authority to suspend or terminate approval of research project that are not being conducted in accordance with the IRB's requirements or that have been associated with unexpected harm to subjects, as well as the responsibility for reporting investigator noncompliance to Health and Human Services' Office for the Protection from Research Risks.

Most institutions where human subject research is conducted have an IRB. It serves primarily as a means of protecting the right of subjects and is an important research quality control measure. The IRB also protects researchers and the institution as well. Among the many other functions of the IRB is the evaluation and approval of proposed studies as well as the informed consent documents that will be used in an investigation.

Components of an Informed Consent

Consent forms may vary considerably in terms of content, language, and length. What follows are descriptions of some of the basic elements of an informed consent document. Check with your university for specific information regarding its consent form requirements.

There are several issues that must be addressed in the written consent documentation. These are as follows:

- Statement that study involves research
- Explanation of the purposes of the research
- Duration of the subject's participation
- Identification of any experimental procedures and/or drugs
- Description of procedures to be followed
- Reasonably foreseeable risks or discomforts
- Direct benefits reasonably expected
- Appropriate alternative procedures or treatments
- Extent of confidentiality of records
- Availability of compensation
- Information regarding availability of medical treatment
- Who to contact with questions about the research project
- Who to contact with questions about rights as a research subject
- Statement that participation is voluntary and refusal to participate will

involve no penalty or loss of benefits otherwise entitled to and that subject may discontinue participation at any time without penalty or loss of benefits

Guidelines for Using Human Subjects in Writing Proposals to the National Institutes of Health

When research involving human subjects the following six points should be addressed

1. Provide a detailed description of the proposed involvement of human subjects in the work previously outlined in the Research Design and Methods section. Describe the characteristics of the subject population, including their anticipated number, age range, and health status. Identify the criteria for inclusion or exclusion of any sub-population. Explain the rationale for the involvement of special classes of subjects, such as fetuses, pregnant women, children prisoners, institutionalized individuals, or others who are likely to be vulnerable.

2. Identify the sources of research material obtained from individually identifiable living human subjects in the form of specimens, records, or data. Indicate whether the material or data will be obtained specifically for research purposes or whether use will be made of existing specimens, records, or data.

3. Describe plans for the recruitment of subjects and the consent procedures to be followed. Include the circumstances under which consent will be sought and obtained, who will seek it, the nature of the information to be provided to prospective subjects, and the method of documenting consent. State if the Institutional Review Board (IRB) has authorized a modification or waiver of the elements of consent or the requirement for documentation of consent. The informed consent form, which must have IRB approval, should be submitted to the PHS only if requested.

4. Describe potential risks (physical, psychological, social, legal, or other) and assess their likelihood and seriousness. Where appropriate, describe alternative treatments and procedures that might be advantageous to the subjects.

5. Describe the procedures for protecting against or minimizing potential risks, including risks to confidentiality, and assess their likely effectiveness. Where appropriate discuss provisions for ensuring necessary medical or professional intervention in the event of adverse effects to the subjects. Also, where

appropriate, describe the provisions for monitoring the data collected to ensure the safety of subjects.

6. Discuss why the risks to subjects are reasonable in relation to the anticipated benefits to subjects and in relation to the importance of the knowledge that may reasonably be expected to result.

Protecting Animal Subjects

Matt (1993) has discussed "Ethical Issues in Animal Research." This is not a new issue, having been discussed in Europe for over 400 years and in the United States for over 100 years. The ground rules were established long ago when Descartes indicated it was justifiable to use animals in research because they could not reason and were therefore "lower" in the order of things than humans.

If animals are used as subjects for research studies in scientific experiments, institutions require adherence to the Guide for the Care and Use of Laboratory Animals, published by the U.S. Department of Health and Human Services, as detailed in the Animal Welfare Act (PL 89-544, PL 91-979, and PL 04-279). Most institutions also support the rules and procedures for recommended care of laboratory animals as outlined by the American Association for Accreditation of Laboratory Animal Care.

Guidelines for Using Animals in Research Proposals to the National Institutes of Health:

The following five points should be addressed:

1. Provide a detailed description of the proposed use of the animals in the work outlined in the Research Design and Methods section. Identify the species, strains, ages, sex, and the numbers of animals to be used in the proposed work.

2. Justify the use of animals, the choice of species, and the numbers to be used. If animals are in short supply, costly, or to be used in large numbers, provide an additional rationale for their selection.

3. Provide information on the veterinary care of the animals involved.

4. Describe the procedures for ensuring that discomfort, distress, pain, and injury will be limited to that which is unavoidable in the conduct of scientifically sound research. Describe the use of analgesic, anesthetic, and tranquilizing drugs and comfortable restraining devices, where appropriate, to minimize discomfort, distress, pain, and injury.

5. Describe any method of euthanasia to be used and the reasons for its selection. State whether this method is consistent with the recommendations of the Panel on Euthanasia of the American Veterinary Medical Association. If not, presents a justification for not following the recommendations.

Institutional Animal Care and Use Committee

As IRB, an Institutional Animal Care and Use Committee (IACUC) is a panel of research experts who pass judgment on the quality and safety of studies before they can be conducted. Most institutions where animal research is conducted have an IACUC. It serves primarily as a means of ensuring animal welfare and is an important research quality control measure.

Through the many steps of research, highly ethical behavior is imperative. While selecting important questions and designing effective research protocols in which research errors are minimized, it is essential to keep in mind that the execution and presentation of the research must be accomplished in an ethical fashion.

Summary

Ethical issues that impact graduate students in their research and scholarly activities were discussed. Misconduct includes fabrication, falsification, plagiarism, or other practices that seriously deviated from those that are commonly accepted within the scientific community for proposing, conducting, or reporting research. Ethics and procedures in the use of human and animal subjects were discussed. The role of Institutional Review Board (IRB) and the components of informed consent form, and the guidelines for using animals in research were also discussed.

70

References

Crase D. Rosato F.D. Single versus multiple authorship in professional journals. J Physical Educ Rec Dance 1992. 63:28-31

Code of ethics for the profession of dietetics. J Am Dietet Assoc 1998;88:1592-1596

Committee on the Conduct of Science, National Academy of Sciences, On Being a Scientist, Washington, D.C: National Academy Press, 1989.

Community of Science: http://www.cos.com/

Council of Biology Editors. Ethics and Policy in Scientific Publication. Bethesda, MD: Council of Biology, Editors, Inc., 1990.

Editorial, Responsible conduct regarding scientific communication. J Neuroscience 1999; 19(1):iii – xvi

Ethics in Science: http://www.chem.vt.edu/ethics/ethics.html

Friedman P.J. Correcting the literature following fradulent publication. JAMA 1990; 263:1416-1419.

Fine M.A., Kurdek L.A. Reflections on determining authorship credit and authorship order on faculty-student collaborations. American Psycologist 1993; 48:1141-1147

Gray, B.H. Human Subjects in Medical Experimentation. New York: John Wiley, 1977.

Greenwald, R.A. Ryan, M.K., Mulvihill, J.E. Human Subjects Rresearch: A Handbook for Institutional Review Boards. New York: Plenum Press, 1982.

Henderson, J. When Scientists Fake It. American Way, 56-62, 100-101, 1990.

Kaiser J. Tulane inquiry clears lead researcher. Science 1999; 284(5422): 1905

Larosa J.C. Tulane investigation completed. Science 1999; 284(5422): 1932

Levine, R.J. Ethics and Regulation of Clinical Research. (2nd ed.), New Haven, Conn: Yale Univ Press, 1988.

Masters K.J. Scientific misconduct. JAMA 1993; 269:3105-3106.

Matt K.S. Ethical issues in animal research. Quest 1993; 45:45-51

National Commission for the Protection of Human Subjects of Biomedical and Behavioral Research. The Belmont Report: Ethical Principles and Guidelines for the Protection of Human Subjects of Research. Washington, D.C.: US Government Printing Office; DHEW publications. (05)78-0012, 78-0013, 78-0014, 1988.

Parrish D.M. Scientific misconduct and correcting the scientific literature. Acad. Med 1999; 74:221-230

Punch, M. The Politics and Ethics of Field-Work. Beverly Hills, CA: Sage, 1986.

Schmitt R.W. Misconduct in science. Science 1993; 262: 1194-1195

Science and ethics: http:mel,lib.mi.us./science/ethics.html

Silverman,W.A. Human Experimentation: A Guided Step into the Unknown. Oxford, England: Oxford Medical Publications, 1986.

The Institutional Guide to DHEW Policy on Protection of Human Subjects, Washington, D.C: Department of Health Education, and Welfare, 1971.

Zelaznik H.N. Ethical issues in conducting and reporting research: A reaction to Kroll, Matt, and Safrit. Quest 1993; 45:62-68

Zuckerman, H. Deviant Behavior and Social Control in Science. Beverly Hills, CA: Sage Publications, 1977.

PART II

==

STATISTICAL AND MEASUREMENT CONCEPTS IN RESEARCH

Part II discusses statistical and measurement concepts in research such as descriptive versus inferential statistics, parametric and non-parametric statistics, and measurement issues associated with dependent variables. Part II is designed for students with no background in statistics and limited backgrounds in mathematics.

Chapter 6 explains the need for statistics. We describe different types of sampling procedures and summarize the basic statistics such as measures of central tendency and measures of variation, and normal distribution.

Chapter 7 presents relationship among variables. We discuss applications of correlation and regression in research.

Chapter 8 covers statistical techniques that are used for describing and finding relationships among variables. They can be used to evaluate the effects of an independent variable on a dependent variable. We discuss student t test, analysis of variance (ANOVA), and analysis of covariance (ANCOVA).

Chapter 9 provides information on non-parametric techniques for data analysis.

Chapter 10 reviews many of the measurement issues that apply when conducting nutrition research. We discuss reliability, validity, systematic and random variability. We also discuss the measurement errors, reliability and validity in dietary assessment.

CHAPTER 6

===

STATISTICAL CONCEPTS

Introduction

Many people incorrectly believe that statistics can be a misleading or manipulation tool that can support a predetermined conclusion. Although statistics may initially seem intimidating, this field plays a pivotal part in the logical interpretation of complex physical process. Statistics can both describe the observed behavior of a subject and support broader conclusions based on the observations. The goal of this chapter is to impart a basic understanding of important terms and methods used in research.

From a math standpoint, the statistical procedures that will be presented involve nothing more complicated than arithmetic. If you can add, subtract, multiply, divide, and find a square and square root with a calculator, you are in business. Performing calculations from an equation is much like following directions on a road map. If one knows what the signs and symbols are that person will have no problem getting to the destination. With equations it's just a matter of performing calculations in order to get to the end result. Nevertheless, as straightforward as this is, it would be very unlikely that a modern researcher would perform statistical computations by hand.

There are numerous statistical software package (Minitab, SPSS, SAS, etc) for mainframes and personal computers that tremendously expedite and simplify the data reduction and analysis processes. It can be argued that once someone understands the basic tenets of a statistic, particularly those related to its function and meaning, hand calculations are of little value. However, some initial number manipulation may be warranted to reach that level of comprehension.

Being a competent user of statistics doesn't mean that one needs to become a professional statistician. Most researchers are users of statistics, not mathematicians or statisticians. They know what tool to select for the job, how to apply it, and how to interpret the results of its use. In everyday life, one can be an effective user of many tools without truly comprehending their underlying mechanics. How many pilots understand the elaborate engineering and mechanical operations of their aircraft? How many bone researchers understand mechanical and engineering operations of a DEXA machine? We effectively operate and maintain sophisticated automobile, but do we really know how they work? Statistical purists may resent this statement but we think it's true: One can effectively use statistics without understanding all of the intricate "engineering" details.

Having said these things the philosophy should be apparent. We will present basic theory and appropriate selection, application, and interpretation of statistical procedures commonly used in health sciences research. It is, however, extremely important to recognize that even apparently minor changes in the design of an experiment may require changes in the statistical analysis. Any time you deviate from a known experimental design (even slightly) or you change the response variable, be sure to talk to a statistician.

Applications of Statistics in Research

Statistics is simply an objective means of interpreting a collection of observations. Various statistical techniques are necessary to allow the description of the characteristics of data, test relationships between sets of data, and test the differences among sets of data. For example, if height and weight were measured in black children in Mississippi, you could sum all the heights and then divide the sum by the number of children. The result (statistics to present the average height) is the mean ($x1 + x2 + x3 + ... =$ sum, Sum $/N =$ mean, where $x1$, $x2$, $x3$ = each child height, N = number of children). The mean (M) describes the average height in the class; it is a single characteristic that represents the data.

The nutrition researchers apply these statistics to compare the mean height between different groups as well as different studies.

Example 6.1

The mean height of boys and girls in Mississippi study was compared with that of children in the Ten-State Nutrition Survey (Figure 1). The mean height of boys and girls in Mississippi study is greater than that of children in comparable age groups in the Ten-State Nutrition Survey, except for 7-year-old boys and 14-year girls except for 7-year-old boys and 14-years girls.

An example of testing relationship between sets of data would be to measure the degree of association between height and age. One measure of the degree of association between two variables (height and age) is called Pearson r (or simple correlation). When two variables are unrelated, their correlation is approximately zero. In Figure 6.1, the two variables (height and age) have a positive correlation. Relationships and correlations are discussed in greater detail in the next chapter, but for now you should see that researchers frequently want to investigate the relationship between variables.

Besides descriptive and correlational techniques, a third category of statistical techniques is used to measure differences among groups. In Table 6.1 shows correlation coefficient with age for selected anthropometric measurements in the same Mississippi study.

Figure 6.1 Comparison of the mean height for the children in the Mississippi survey with the mean heights for black and white children in the Ten-State Nutrition Survey (6), by age and sex.

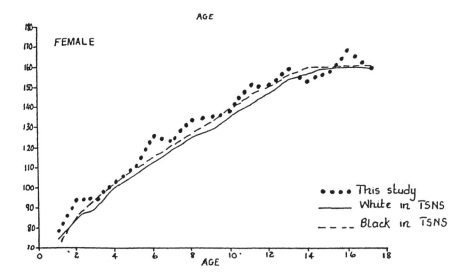

Figure 6.1 (continued)

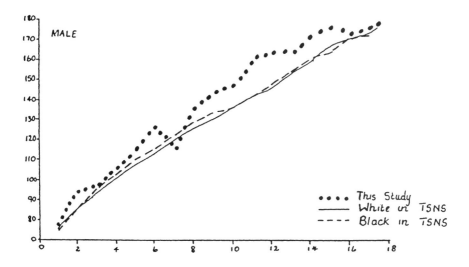

Table 6.1

MEAN LEVELS OF SELECTED ANTHROPOMETRIC MEASUREMENTS BY AGE AND SEX FOR ALL
PERSONS OVER 5 YEARS OLD

Age Group (yr.)	Anthropometric Measurements				
	Diastolic Pressure mm. Hg	Subscapular Skinfold mm.	Triceps Skinfold mm.	Weight kg.	Height cm.
Female Subjects					
6-11 (35)*	67.4 ± 2.7	11.1 ± 1.3	12.7 ± 1.5	32.0 ± 2.0	138.5 ± 4.7
12-17 (45)	70.3 ± 2.2	14.0 ± 1.1	14.1 ± 1.0	55.6 ± 2.7	160.6 ± 4.0
18-34 (60)	78.4 ± 2.0	23.6 ± 1.4	22.5 ± 1.3	66.0 ± 2.1	161.8 ± 2.8
35-60 (36)	92.1 ± 3.2	31.7 ± 2.1	26.0 ± 1.8	77.3 ± 4.2	163.2 ± 4.7
Over 60 (13)	79.8 ± 7.5	29.0 ± 3.9	28.9 ± 3.2	75.2 ± 7.1	159.8 ± 12.9
Total Mean	77.2 ± 1.1	20.8 ± 0.9	19.8 ± 0.7	60.2 ± 1.6	157.4 ± 1.2
Male Subjects					
6-11 (21)	71.0 ± 4.3	12.5 ± 3.2	8.8 ± 1.3	32.7 ± 3.0	139.6 ± 7.8
12-17 (65)	69.7 ± 1.6	9.5 ± 0.5	10.6 ± 0.6	51.0 ± 1.7	165.5 ± 3.0
18-34 (12)	77.7 ± 7.2	17.4 ± 2.2	14.1 ± 1.9	76.7 ± 7.6	174.2 ± 15.4
35-60 (8)	85.7 ± 12.6	21.8 ± 2.6	15.8 ± 2.6	75.8 ± 10.7	172.0 ± 23.1
Over 60 (8)	94.0 ± 13.6	24.3 ± 5.9	16.8 ± 4.4	76.1 ± 11.4	168.0 ± 22.6
Total Mean	73.7 ± 1.3	12.9 ± 0.9	11.5 ± 0.6	53.9 ± 1.8	162.2 ± 2.5
Correlation Coefficient With Age	0.599†	0.550†	0.489†	0.605†	0.288†

* Sample Size
* Mean ± Standard Error
† P < 0.001

Levels of measurements

Levels of measurement allow us to know something about the intricacy of the numbers or data that we are generating from our research. It is important to know the level of measurement because it directly affects our choice of the correct statistical procedure.

Nominal Level

Nominal level data are the most rudimentary because they only provide information which "names" or "categories" an individual. In a sport uniform number 15 is not the same as number 10, similarly interstate highway number 80 is different than number 29. These are examples of nominal values. Nominal levels are also associated with counted categorical data. Categories need to be defined in such a way that they are mutually exclusive; and an observation must belong to one and only classification. For instance, you could categorize cuisine as French, Italian, Chinese, Korean, or the like. The categories are mutually exclusive, because cuisine can't be classified as both a French cuisine and an Italian cuisine or. Usually nominal level data is summarized by giving the number and percent in each category.

Ordinal Scale

Ordinal level data demonstrate ranks, that is, they allow us to know when one individual measurement is "greater than" or "less than" another individual measurements. Thus they provide more information than nominal data. Someone who finishes second in a road race did better than someone who was third, and of course, third place was better than fourth. What we don't know from rank data is the amount of difference between each because there is no equality of units. For instance, the difference between second and third place may have been 0.5 second and the difference between third and fourth may have been 30.0 seconds. Note that ordinary arithmetic operations (addition, subtraction, multiplication, division) do not make sense when performed on measurements at the nominal or ordinal level.

Interval Level

Interval level data have all of the properties of ordinal data plus equality of units. On a Celsius temperature scale we know that $25°$ is as warmer than $20°$ as $20°$ is warmer than $15°$. Interval level data inform us about difference, direction of difference, and amount of difference in equal units.

Ratio Levels

Ratio level data are the most complex because they have all of the elements of interval level, plus they have an absolute zero point. What is an absolute zero? No, not -273° C, or at least not exactly. An absolute zero point represents the absence of the trait being measured. Variables such as weight, distance, and elapsed time have meaningful absolute zero points. Interval level zeros are not absolute. For example, if it is 0°C is there no temperature? Sure there is, and it is quite cold. If someone can't do at least one pull-up, does that mean she has no upper body strength? Of course not. Absolute zero points also allow us to make comparison like 5 inches is half the distance of 10 inches or 60 seconds is three times as long as 20 seconds. Without an absolute zero point such comparisons are not valid. Does something that is 60°C have three times as much heat energy as something that is 20°C? The object is 40° warmer but, not three times (300%) as warm. The reason is that the zero point on the Celsius scale is not an absolute zero. Because the true absolute zero point for temperature is -273°C, then 20° = (273 + 20 =293) and 60° = (273 + 60 =333). Therefore, 333/293 = 1.1365 - or 60° is only 13.65% warmer than 20°!

Descriptive and Inferential Statistics

Descriptive statistics are employed to measure a trait or characteristic of a group without generalizing that statistic beyond that group. The range of your class test scores on the last research method, your grade point average, and grade point average of the freshmen are all examples of descriptive statistics.

Inferential statistics are used to make generalizations or inferences from a smaller group to a larger group. An inference or a generalization is a type of prediction that is made by measuring a trait from a smaller representative group and estimating what it would be in the larger group. In order to make valid inferences, it is critical that the subjects in the smaller group are good representatives of the larger group. For example, a nutritionist is interested in determining the cholesterol levels of the 10,000 students on a university campus so that they can be compared to national norms. Does the researcher need to test all of the students to answer this question? The answer is "No." This is because of the power to infer with statistics. The researcher obtains a group of 100 subjects who are a good model of the entire student body. How this can be done is discussed in the next section, on sampling. The researcher now tests the subjects and determines their cholesterol levels. Because the 100 subjects are representative of the entire student body, an inference may be made to estimate (usually with good accuracy) what the typical cholesterol levels of all 10,000 students would be. From this example, it should be needless to say that inferential statistics are one of the most valued tools of a researcher. The method of selecting the sample, procedures, and context is what does or does not allow inference.

The sample is the group of subjects, treatments and situations on which the study is conducted. The key issue is how these samples are selected. In the following sections we discuss the types of sampling typically used in designing studies and statistical analysis.

SAMPLE SELECTION

Random Selection

The sample of subjects might be randomly selected from some larger group, or a population. For example, if your university 10,000 students, you could randomly select 200 for a study. You would assign each of the 10,000 students a number-name. The first number-name would be 0000, the second 0001, the third 0002, up through the last (the 10,000th), who would have the number-name 9999. Then a random numbers table (see Table A.1 in Appendix A) would be used. The numbers in this table are arrange in two-digit sets so that any combination of row or column is unrelated. In this instance (number-name 0000-9999), you need to select 200 four-digit numbers. Because the rows and columns are unrelated, you can choose any type of systematic strategy to go through the table. Enter the table at random. Suppose the place of entering the table is the eighth column of a two-digit number on row 10. The number-name here is 2334. You select the subject with that number-name (2334), then 6394, and so on until you have selected 200 subjects.

The system used in the random number table is not the only one. You can use any systemic way of going through the table. You could read across rows rather than down columns. Of course, the purpose of all this is to select a sample of subjects randomly so that the sample represents the larger population; that is the findings in the sample can be inferred back to the larger population. From a statistical view point, this says that a characteristic, relationship, or difference found in the sample is likely also to be present in the population from which the sample was selected (inference).

Stratified Random Sampling

In stratified random sampling, the population is divided (stratified) on some characteristic before random selection of the sample. For example, the selection was of 200 subjects from a population of 10,000 in four different counties. Suppose 30% of the population is in Cleveland County, 30% is in Jefferson county, 20% is in Kennedy county, and 20% is in Oklahoma county. You could stratify on county before random selection to make sure the sample was exact in terms of county representation. Here you would randomly select 60 subjects from the 3,000 in Cleveland county, 60 from the 3,000 Jefferson county, 40 from the 2,000 Kennedy county, and 40 from the 2,000 in Oklahoma county. This still yields a total sample of 200. Stratified random sampling might be particularly appropriate for survey or interview research.

Systemic Sampling

If the population from which the sample is to be selected is very large, assigning a number-name to each potential subject is time consuming. Suppose you want to sample a town with a population 100,000 for cancer study. One approach would be to use systematic sampling from the telephone book. You might decide to call a sample of 1,000 people. To do so, you would select every 100th name in the phone book (100,000/1,000 = 100). Of course, you are assuming that the telephone book represents the population and that everyone you need to sample has a listed telephone number. The Framingham study used this approach.

Random Assignment

In experimental research, groups are formed within the sample. The issue here is not how the sample is selected but how the groups are formed within the sample. The investigator should not allow the subjects to select what group they would like to be in as this represents a form of selection bias that weakens the experiment. Random numbers, blind draws of coded cards, and other such unbiased methods may be used. Animal experiments use this technique. In all animal studies, the groups within the animals be randomly assigned or randomized. It is also called "randomized block assignment."

Central Tendency

Some of the more easily understood statistical and mathematical calculations are those that central tendency and validity of scores. When you have a group of scores, one number may be used to represent the group. The number is generally the mean, median, or mode. These terms are ways of expressing central tendency. Within the group of scores, each individual score will differ to a given degree from the central tendency score. The degree of difference is the score's variability. Terms that describe the variability of the scores are **standard deviation and variance, and the difference between the quantities.**

The mean

When people use the word "average," they are usually referring to the arithmetic mean. If you have totaled the test grades you have made in a subject during a school term and divided by the number of tests taken, you have computed the arithmetic mean. The arithmetic mean is the most commonly used average.

$75 + 75 + 65 + 50 = 265 \quad 265/4=66.25$

The computations may show characteristics of a population (parameters), or they may produce results taken from a sample (statistics). Statisticians maintain the distinction between parameters and statistics through the use of different symbols. Greek letters are generally used to denote parameters, while lower case italic letters denote sample statistics. Table 6.2 shows some common symbols. As you can see, a population mean or percentage is designated by μ (mu) or π (pi), while a sample mean or percentage is denoted by X or p. Similarly, the standard deviation is identified by σ (sigma) if it is computed from a population and by s if sample data are used.

Table 6.2   Distinctions between a population and a sample

Area of Distinction	Population	Sample
Definition	Defined as a total of the items under consideration by the researcher	Defined as a portion of the population selected for the study
Characteristics	Characteristics of a population are parameters	Characteristics of a sample are statistics
Symbols	Greek letters or capitals: μ = population mean σ = population standard deviation N = population size π = population percentage	Lower case italic letters: $\overline{X}$ = sample mean s = sample standard deviation n = sample size p = sample percentage

$$\mu = \frac{\Sigma x}{N}$$

$$\overline{X} = \frac{\Sigma x}{n}$$

Where x = the values of a variable
μ = arithmetic mean of a population
$\overline{X}$ = arithmetic mean of sample
Σ = "the sum of"
N = number of x items in the population
n = number of x items in the sample

The Median

The median is a measure of central tendency that occupies the middle position in an array of values. That is, half the data items fall below the median, and half are above that value. Note that the word "array" has been emphasized; it is necessary to put the data into an ascending or descending order before selecting the median value.

For example:

n = odd number   (n + 1)/2 = (7 + 1)/2 = 4, in which 4th number is the median position
10, 9, 6, 5, 3, 2, 1

The middle value (median) in the array is 5.

n = even number
10, 9, 6, 6, 5, 3, 2, 1

The middle position in the array would be the mean of the two middle scores 6 and 5 --- i.e., the median value would then be 5.5.

The mode

The mode, by definition, is the most commonly occurring value in a series. The mode of the following series is 6.

10, 9, 6, 6, 6, 5, 3, 2, 1

The Variability Score; The standard Deviation

The standard deviation is also used with the mean, and it is generally the most important and useful measure of dispersion. In a precise sense, the population standard deviation is the square root of the average of the squared deviations of the individual data items about their mean. The standard deviation is an estimate of how far away items in a data set are from their population mean.

Before we can compute a standard deviation, we must determine if our data set represents a sample. To calculate the standard deviation for a sample data set is used, the standard deviation is found with

$$s = \sqrt{\frac{\Sigma (x - X)^2}{n - 1}}$$ where s = sample standard deviation
x = values of the observations
X = mean of the sample
n = number of observations in the sample

It is not too difficult to use the formula with a small sample data set. But when many more items are included, the procedure becomes tedious. In that case, you might prefer to use a "shortcut" variation formula. It can be shown algebraically that the following formula is equivalent to the previous one:

$$s = \sqrt{\frac{\Sigma x^2 - 1/n (\Sigma x)^2}{n-1}}$$

One final point for later consideration is that the square of the standard deviation is called the variance, or s^2.

The Range Score

The range measures just when its name implies: the distance between the endpoints in a distribution. The range is determined by taking the difference between the high and low scores in a distribution, with some statisticians advocating adding 1 to that difference. Here is an example of a range calculation:

Range = High score - Low score or
Range = (High score - Low score) + 1

Data: 3,4,5,6,7,8,9
Range = 9 - 3 = 6 or (9 - 3) + 1 = 7

The range is a weak measure of variability because it is based on the extreme scores in a distribution and determined by only two scores. The range scores are often presented in students' theses or dissertations.

Two Categories of Statistical Tests: Parametric and Nonparametric Statistics

Statistical analyses can be classified as parametric or nonparametric types. Parametric statistics are used when data are interval or ratio level and when the populations from which the observations were made are thought to be normally distributed.

The parametric statistical tests have three assumptions about the distribution of the data:
1. The population from which the sample is drawn must be normally distributed on the variable of interest.
2. The populations from which the samples are drawn have the same variance.
3. The observations are independent.

Nonparametric statistics are used when data are nominal or ordinal level or when the populations from which the observations were made are thought to be nonnormally distributed. Sometimes these statistics are referred to as distribution-free analyses. We will discuss in detail in the chapter 9.

When the assumptions for these use are met, parametric statistics are more powerful than their nonparametric counterparts because they are more likely to reject a false null hypothesis. However, when the criteria for the use of a parametric analysis cannot be met, a nonparametric alternative is more powerful and should be used.

To have power means to increase the chances of rejecting a false null hypothesis. You frequently assume that the three assumptions for use of parametric statistics are met. The assumptions can be tested by using estimates of skewness and kurtosis. Remember that the assumptions are with respect to the population, not the sample.

The Standard Normal Distribution

To understand skewness and kurtosis, first consider the normal distribution in Figure 6.2. This is a normal distribution curve that is characterized by:

1. It is bell-shaped.
2. The mean, median, and mode are equal and located at the center of the distribution (Figure 6.3).
3. It is unimodal (i.e., it has only one mode)
4. The curve is symmetrical about the mean, which is equivalent to saying that its shape is the same on both sides of a vertical line passing through the center.
5. The curve is continuous --- i.e., there are no gaps or holes. For each value of x, there is a corresponding value of y.

Figure 6.2 Standard Normal Distribution Curve

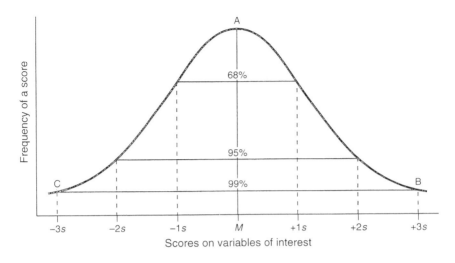

Figure 6.3 Mean, Median and Mode are Equal and Located at the Center of the Distribution

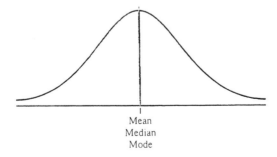

6. The curve never touches the x-axis. Theoretically, no matter how far in either direction the curve extends, it never meets the x-axis but gets increasingly closer.
7. The total area under the normal distribution curve is equal to 1.00, or 100%. This feature may seem unusual, since the curve never touches the x-axis, but this fact can be proven mathematically by using calculus.
8. The area under the normal curve that lies within one standard deviation of the mean is approximately 0.68, or 68%; within two standard deviations, about 0.95, or 95%; and within three standard deviations, about 0.997, or 99.7%.

You must know these properties in order to solve problems using applications of the normal distributions. If the data values are evenly distributed about the mean of the distribution, it is said to be a symmetrical distribution. When more of the data values fall to the left or to the right of the mean, the distribution is said to be skewed. Figure 6.4 shows a negatively skewed distribution; the majority of the data values fall to the right of the mean. Figure 6.5 shows a positively skewed distribution; the majority of data values fall to the left of the mean. The mean is to the right of the median, and the mode is to the left of the median. The "tail" of the curve indicates the direction of skewness (right: positive; left: negative).

Figure 6.4 Negatively skewed   Figure 6.5 Positively skewed

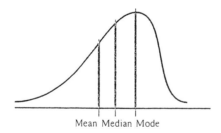
Mean Median Mode

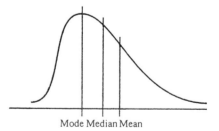
Mode Median Mean

The curves with abnormal kurtosis are shown in Figure 6.6 with more peaked shape and in Figure 6.7 with more flat curve than normal curve. They have different standard deviations with either same or different means between the two curves.

Figure 6.6 More peaked (leptokurtic)   Figure 6.7 More flat (platokurtic)

The standard normal distribution is shown in Figure 6.8. The values under the curve indicate the proportion of area in each section. For example, the area between the mean and one standard deviation is about 0.3413, or 34.13%.

Figure 6.8 Normal distribution

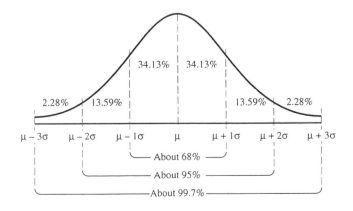

Application of the Normal Curve

Let us say that a researcher has measured the percent of body fat of 2,000 women. The resulting distribution has a mean = 25.0% fat and a standard deviation of 4.0. Therefore, the scores would be distributed in the following manner:

68% = 25.0 ± 1 (4.0) = 21.0 to 29.0 % body fat
95% = 25.0 ± 2 (4.0) = 17.0 to 33.0 % body fat
99% = 25.0 ± 3 (4.0) = 13.0 to 37.0 % body fat

The z Score

The z score is a statistic that is used to convert a raw score from a distribution into units of the normal curve called standard deviation units. (The unit normal curve is sometimes referred to as the z distribution.) A z score provides information about how far a raw score is away from the mean, in standard deviation (SD) units. The equation for the z score is:

$$z = \frac{value - mean}{standard\ deviation} \quad or \quad z = \frac{x - \mu}{\sigma}$$

For instance, the mean value is 25.0 and SD is 4.0, from the women's body fat example, what would be the z score for 17 and 37% body fat?

$$z = \frac{17 - 25}{4} = -2.0 \qquad z = \frac{37 - 25}{4} = 3.0$$

A positive z score tells you how many SD units a score is above the mean, whereas a negative score is interpreted as the number of units below the mean. A z score of -2.0 means that a score of 17% body fat is 2.0 SD units below the mean of 25 and a z score of 3.0 means that a body fat of 37% is 3.0 SD units above the mean. Thus, the z score informs you about the exact position of the raw score in a distribution. Because about 99% of all scores in a distribution are between ± 3 SD, then 99% of all z scores will range between -3.0 and +3.0. Look at z distribution in Table A.2 in the appendix to confirm this fact. You will get 99.73% for z score of 3.0.

Appendix A contains Table A.2 is a unit normal distribution (z) for a normal curve. The column z shows the location of the mean. When the mean is the center of the distribution, its z is equal to .00; thus, .50 (50%) of the distribution is beyond the mean, leaving .50 (50%) of the distribution as a remainder.

As the mean of the distribution moves to the right in a normal curve (say to a z of + 1s), .8413 (84%) of the distribution is to the left of the mean (remainder) and .1587 is to the right of the mean (beyond). This table allows you to determine the percentage of the normal distribution included by the mean plus any fraction of a Suppose you want to know what percentage of the distribution would be included by the mean plus one half (.50) of a standard deviation. Using Table A.2 you can see that it would be .6915 (remainder), or 69%.

The z score may also be used to compare from different distributions. Say that a senior female student has body weight of 55 kg and height of 165 cm. Without any other statistical information, however, we would not know how her body weight and height stand as compared to her classmates. Given the following information, we may convert the scores into the same unit of measure (that is, SD units) by computing a z score for each and then make comparisons:

Class mean weight = 50.0 kg SD = 5.0; and class mean height = 172.0 cm SD = 12

$$\text{Body weight } z = \frac{55 - 50}{5} = 1.0 \qquad \text{Height } z = \frac{172 - 165}{12} = -0.5833$$

With the z score information, it can be concluded that body weight of 55 kg was 1.0 SD units above the mean and was heavier than the average, and height of 165 cm was - 0.58 SD units below the mean and was smaller than the average.

Probability

Another concept that deals with statistical techniques is probability, which asks what are the odds that certain things will happen. You use probability in everyday events: "There is a 40 percent chance that it will rain today." "The odds are 75 to 1 of the Baltimore Oriole winning the World Series." "The potential of winning the lottery is 1 in 10,000." "A male between the ages of 55 and 65 has 1 in 100 chances of dying from heart disease." All of these are probability statements that imply the likelihood or chance of a certain event occurring.

Probability plays an informational role in many aspects of our lives.

Statistical decisions are not exact but are made with a certain probability or chance of being right or wrong. Therefore, a fundamental knowledge of probability is requisite for understanding many statistical tests.

In a statistical test, you sample from a population of subjects and events. You use probability statements to describe the confidence you place in the statistical findings. Frequently, you will encounter a statistical test followed by a probability statement such as $p < 0.05$. This interpretation would be that a difference or relationship of this size would be expected less than 5 times in 100 as a result of chance.

The Normal Distribution Curve as a Probability Curve

The normal distribution curve can be used as a probability distribution curve for normally distributed variables. Recall that the normal distribution is a continuous distribution, which means that there are no gaps in the curve (Figure 6.2: normal curve characteristics). In other words, for every z value on the x-axis, there is a corresponding height or frequency value (y axis). Therefore, we can estimate the area under the curve. The area under the curve is more important than the frequencies. This area under the curve corresponds to a probability.

For instance, the critical values are ±1.96 and are placed on each end of the curve (Figure 6.8). In this case that means that the area under the curve from the mean of zero to 1.96 is about 47.5%, with the remaining 2.5% outside that value. This is also true for -1.96. Therefore, 95% of the sampling distribution values are between ±1.96 and 5% of the values are outside those values, with 2.5% on each tail of the curve.

Example 6.2

A nutritionist is interested in determining the incidence of hypercholesterolemia of white men between ages of 50 and 70 years. The prevalence of hypercholesterolemia is of interest to her for the programming and staffing of risk reduction programs. She obtains a random sample of 300 white men and measures their blood cholesterol and finds the mean (M) = 259.0 mg/dL with a standard deviation (SD) = 15.0. This distribution would be divided accordingly:

68% = 259.0 ± 1 (15.0) = 244 to 274 mg/dL
95% = 259.0 ± 2 (15.0) = 229 to 289 mg/dL
99% = 259.0 ± 3 (15.0) = 214 to 304 mg/dL

From this example, what would be the probability that if any one subject was selected at random, he would have a cholesterol level between 244 and 274 mg/dL? Because 68% of the distribution has a plasma cholesterol level between those values, then 68% of 300 subjects = 300 x (0.68) = 204.
Therefore, the probability is p = 204/300 = 0.68

What would be the probability that if any one subject was selected at random, he would have a blood cholesterol level above 274 mg/dL? Because 16% of the distribution is above 1 SD from the mean that represent 48 subjects [300 x (.16) = 48], therefore, the probability is p = 48/300 = .16. What would be the probability that if any one subject was selected at random, he would have a blood cholesterol below 214 mg/dL? Because only 0.5% of the distribution is below this level, then p = 1.5/300 = 0.005.

Actually these problems could be solved easily by computing z scores, referring to the z distribution table, and reading a probability directly from it. For instance, from the sample of 300 men, the z score for a blood cholesterol of 244 mg/dL would be z = 244 - 259/15 = - 1.0, whereas a z score for 274 mg/dL is z = 274 - 259/15 = 1.0. By inspecting the z score table (A.2 table in Appendix), we determine that 68.27% of the area under the curve is between z scores ±1.0. So the probability of any of 300 men having a blood cholesterol between 244 and 274 mg/dL is p = .68 or 68%.

Hypothesis Testing

We defined research hypothesis as a scientific hunch that an investigator has about the expected outcome of a study. Decisions to accept or reject the hypothesis must be based on an objective and logical statistical process generally called hypothesis testing. The hypothesis test is a strict statistical operation that is built on making probability statements for two possible states of reality.

Null and Alternative Hypotheses

The two states of reality that the hypothesis test is predicated on are the null and alternative hypotheses. Sometimes these are referred to as statistical hypotheses because statistical procedures are used to estimate whether they are true or not. The researcher uses the results of the hypothesis test to draw conclusions about the validity of the research hypothesis.

The null hypothesis (H_0) is traditionally defined as a statement of no difference or no relationship. This hypothesis may be written numerous ways, depending on the research condition and the statistical test. For instance, the null hypothesis for comparing two population means (μ) would be written as H_0: $\mu_1 - \mu_2 = 0$ or $\mu_1 = \mu_2$. Both of these statements say the same thing; there is no difference between the two population means.

The alternative hypothesis (H_A) is the logical state of reality that must exist if the null hypothesis is not true. In case of comparing two different population means, the null hypothesis stated that there was no difference between them. If this is not true, then what has to be true? There is a difference between them. The alternative hypothesis in this case is written as H_A: $\mu_1 - \mu_2 \neq 0$ or $\mu_1 \neq \mu_2$, both of which say that there is a difference between population means.

Two-Tailed versus One-Tailed Hypothesis Tests

When a researcher cannot hypothesize about the direction of the outcome of a study, a two-tailed hypothesis test must be used. For example, when comparing two different methods for lowering cholesterol, the researcher does not know which method is going to be better than the other. Therefore, he/she will use two-tailed hypothesis test.

Alpha

In research, the test statistic is compared a probability table for that statistic, which tells you what the chance occurrence is. The experimenter may establish an acceptable level of chance occurrence (called alpha, α) before the study. This level of chance occurrence can vary from low to high but can never be eliminated. For any given study, the probability of the findings being due to chance always exists.

In biomedical research, alpha (probability of chance occurrence) is frequently set at 0.05 or 0.01 (the odds that the findings are due to chance are either 5 in 100 or 1 in 100). There is nothing magical about 0.05 or 0.01. They are used to control for a Type I error. In a study, the experimenter may make two types of error.

A Type I error is to reject the null hypothesis when the null hypothesis is true. For example, a researcher concludes that there is a different effect on bone mass between calcium supplement and non-supplement groups, but there really is not. A Type II error is not reject the null hypothesis when the null hypothesis is false. For example, a researcher may conclude there is no difference between the two dietary treatments, but there really is a difference.

Table 6.3 is called a truth table, which displays Type I and Type II errors. As you can see, to accept a true null hypothesis or reject a false one is the correct decision. You control for Type I errors by setting alpha. For example, if alpha is set at 0.05, then if 100 experiments are conducted, a true null hypothesis of no difference or no relationship would be rejected on only 5 occasions. Although the chances for error still exist, the experimenter has specified them exactly by establishing alpha before the study.

Table 6.3: Truth table for null hypothesis (H_0)

	H_0 true	H_0 false
Accept	Correct Decision	Type II error (β)
Reject	Type I error (α)	Correct Decision

Beta

Although the magnitude of Type I error is specified by alpha, you may also make a Type II error, the magnitude of which is determined by beta (β). Lowering the alpha level to a more rigorous quantity reduces the chance for making a Type I error, but simultaneously increases the probability of making a Type II error. Table 6.4 illustrates the effect that raising and lowering the alpha above and below 0.05 has on Type I and II errors.

So if it is not possible to have an equally low probability of making both errors, which one is more important not to make? The answer to this question is based on the type of research problem the investigator is dealing with. The alpha level 0.05 is mainly used in most of the scientific research.

Table 6.4:   Effect of alpha and probability errors

alpha	Type I Error	Type II Error
0.01	Decreased	Increased
0.05	---	---
0.10	Increased	Decreased

Meaningfulness (Effect Size)

In addition to reporting the significance of the findings, scholars need to be concerned about the meaningfulness of the outcomes of their research. The meaningfulness of a difference between the two means can be estimated in many ways, but the one that has gained the most attention recently is effect size (ES). The formula for ES is:

$ES = (M_1 - M_2) / s$

For health research: 0.2 or less is a small ES; about 0.5 is a moderate ES; and 0.8 or more is a large ES.

However, meaningfulness is usually associated with clinical significance, whereas effect size is more often used for sample size calculation.

Power (1-β)

Power is the probability of rejecting the null hypothesis when the null hypothesis is false (e.g., detecting a real difference), or the probability of making a correct decision. Having power in the statistical analysis is important because it increases the odds of rejecting a false null hypothesis.

As β decreases, power increases. The investigator specifies β during the planning of the study, which then determines power and sample size. For example, if the risk of missing a difference is set at 20% (p=0.20), the chance that the study would find a real difference would be 80% (p=0.80).

Determining Required Sample Size

Research design often requires a compromise of the ideal and feasible; the goal of design is a practical and economic balance between power and sample size. However, in deciding to compromise power, the investigation must recognize that the ability of the study to accomplish its objective is also compromised. A study may not be worth doing if there is a low probability of detecting a meaningful effect with the available sample size.

Sample size determination is critical because it establishes both the reliability of the statistics and the size of the effort. By setting several key research parameters in the planning phase, the researcher can estimate a reasonable sample size. These parameters include the desired significance level, the nature of the statistics to be complied, and estimate of the difference and standard deviation of certain variables.

The probability that, if a true effect exists, a study will detect it (i.e. power) largely depends on the sample size. Increasing the sample size increases the power. At the same time, increasing sample size decreases the risk of a false negative conclusion (i.e. β or type II error) because the ability to detect a true difference is increased.

Example 6.3: Paired Observations

An example might be a study to assess whether a particular intervention will decrease plasma cholesterol by dietary treatment. For example, before and after treatment, and experimental effect can be tested with a paired t test.

The power of a study also depends to some degree on the true magnitude of difference or effect under the study. For any given power, a large difference can be detected with a smaller sample size than can a small difference. The investigator determines in advance the magnitude of difference or effect that is important for the study to be able to detect.

The general relationship among sample size, power, and magnitude of the difference or effect sought can be expressed as:

$$\text{Sample size } (n) > 2 \left(\frac{(\alpha \text{ error} + \beta \text{ error}) \times SD}{|\text{Difference}|} \right)^2$$

Where mathematically the α and β errors are converted to the standardized normal deviates (z values) for the probabilities, the SD equals the estimated standard deviation, and the difference is the absolute value of its magnitude. For instance, a change in plasma cholesterol levels from 225 mg/dl to 180 mg/dl would be a difference of -45 mg/dl or an absolute difference of 45 mg/dl.

In advance, α is specified as 0.05 with a two tailed test. The decision is made to set power or $1 - \beta$ at 0.8, making $\beta = 0.20$. The sample size needed to conduct a test with significance-level α and power, $1 - \beta$, is:

$$n = \left(\frac{(z_{1-\beta} + z_{1-\alpha}) \times SD_{diff}}{|\mu 1 - \mu 0|} \right)^2$$

The quantities $z_{1-\beta}$ and $z_{1-\alpha}$ are values from the standard normal distribution analogous to α and β. Appendix Table A.2 gives selected values of $z_{1-\beta}$ and $z_{1-\alpha}$ corresponding to commonly used values of α and β. Cholesterol change of 45 mg/dl, and SD of difference 50mg/dl.

$$n = \left(\frac{(z_{1-\beta} + z_{1-\alpha}) \times SD_{diff}}{|\mu_1 - \mu_0|} \right)^2$$

$$= \left(\frac{(0.84 + 1.96) \times 50}{45} \right)^2$$

$$= 9.67 \text{ or } 10 \text{ subjects}$$

Example 6.4: Stratified Random Sampling

Stratified random sampling presents several issues in regard to sample size. Investigative studies involving a single dichotomous stratification parameter (urban vs. rural; smoker vs. nonsmoker, alcohol drinker vs. non-drinker, etc.) With random sampling in each stratum may employ a formula to determine sample size. The formula considers confidence level and sampling error in calculating a representative sample size.

$$n = (z/e)^2 (p)(1-p)$$

where n = sample size
  z = the standard score corresponding to a given confidence level
  e = the portion of sampling error in a given situation
  p = the estimated proportion or incidence of cases in the population

Confidence level indicates the probability that sample proportion will reflect the population proportion with a specific degree of accuracy (sampling error designated as e in the formula). With a 95% confidence level z = 1.96. Suppose that a nutritionist decided to investigate nutrition education programs in male and female adults. In ascertaining the sampling frame, it is 40% males. If p = .4, z = 1.96, and sampling error = 0.1, then the following calculations apply:

$$n = (1.96/0.10)^2 (0.4) (0.6) = (19.6)^2 (0.4) (0.6) = 92.19$$

sample size = 93

Surveys require a greater sample size than experimental studies because of response failure, item omission, poor interviewing, and so on.

Summary

In this chapter we have tried to make the point that the type of statistics used does not determine whether findings can be generalized; rather, it is sampling that permits inference. Whenever, possible, random sampling is the method of choice. In type of surveys, stratified random sampling is desirable for the study to represent certain segments of a population. In experimental research, random assignment of subject to the groups is definitely desirable so that the researcher can assume equivalence at the beginning of the experiment.

We began the coverage of statistical techniques with basic concepts such as measures of central tendency and variability and normal distribution. It is important to remember that statistics can do two things: establish significance and assess meaningfulness.

Probability is an important component of statistics. Probability statements refer to the confidence you place in the statistical findings. The null hypothesis is used in statistical tests. It states that there is no difference among the experimental groups and that any observed findings is simply a chance occurrence.

The possibility of committing statistical error always exists. A Type I error is rejecting the null hypothesis when it is true. A Type II error is accepting the null hypothesis when it is false. These two errors work in opposition: If you try to reduce Type I error, then you increase Type II error. The researcher must decide what level of significance (alpha) to establish.

The most important issue in statistics is using good judgment. This means the concept of power ---- estimating the characteristics needed to determine whether the outcome of a study have merit: the combination of alpha, effect size, power (1- beta), and sample size. Research design often requires a compromise of the ideal and feasible; the goal of design is a practical and economic balance between power and sample size.

References

Armitage, P. Statistical Methods in Medical Research. London: Blackwell Scientific Publications, 1971.

Cheny, C.L. Boushey, C.J. Estimating Sample Size. In Research - Successful Approaches. Am Dietet Assoc, 1992.

Cohen, J. Statistical Power Analysis for the Behavioral Sciences, (2nd ed.) New York: Academic Press, 1969.

Cohen N.L., Laus M.J., Stuzman N.C., Swicker R.C. Dietary change in participants of the better eating for better health course. J Am Dietet A1991; 91: 345-349

Conocer, W.J. Pratical Nonparametric Statistics. New York: Wiley, 1971.

Dixon, W.J., Massey, F.J. Jr. Introduction to Statistical Analysis. (3rd ed.) New York: McGraw-Hill Book, 1969.

Donner A., Eliasziw M. Sample size requirements for reliability studies. Stat Med 1987; 6:441-445

Dupont W.D., Plummer W.D., Jr. Power and sample size calculations: a review and computer program. Controlled Clin Trials 1990; 11: 116 – 124

Fleiss J.L., Tytun A., Ury H.K. A simple approximation for calculating sample sizes for comparing independent proportions. Biometrics 1980; 36:343-346

Frank B.D., Huck S.W. Why does everyone use the .05 significant level? Research Quarterly Exercise Sport 1986; 57:245-249

Ireton-Jones, C.S., Gottschlich, M.M., Bell, S.J. Practice-Oriented Nutrition Research: An Outcomes Measurement Approach. Gaithersburg, MD: Aspen, 1998.

Koh E.T. Selected anthropometric measurements for low income, black population in Mississippi. J Am Dietet A 1981; 79:555-561

Lachin J.M. Introduction to sample size determination and power analysis for clinical trials. Controlled Clin Trials 1981; 2:93-96

Lehmer E. Inverse tables of probabilities of errors of the second kind. Annal Mathemat Stat 1994; 15: 388-398

Thomas, J.R., Nelson, J.K. Research Methods in Physical Activities. (3rd ed.) Champaign, IL: Human Kinetics, 1996.

Van Horn, L., Moag-Stahlberg A., Liu K. Effects of serum lipids of adding instant oats to usual American diets. Am J Public Health 1991; 81:183-186

CHAPTER 7

===

RELATIONSHIPS AMONG VARIABLES

Introduction

In the previous chapter hypothesis testing was explained. Another area of inferential statistics involves determining whether a relationship between two or more variables exists. For example, nutritionists may want to know whether caffeine intake is related to heart disease, or whether person's age is related to her blood pressure. A zoologist may want to know whether the birth weight of a certain animal is related to the life span of the animal. These are only a few of the many questions that can be answered by using the technique of correlation and regression analysis. Correlation is a statistical method used to determine whether a relationship between variables exists. Regression is a statistical method used to describe the nature of the relationship between variables --- i.e., a positive or negative, linear or nonlinear relationship.

The purpose of this chapter is to answer the following questions statistically:

1. Are two or more variables related?
2. If so, what is the strength of the relationship?
3. What type of relationship exists?
4. What kind of predictions can be made from the relationship?

In order to answer the first two questions, statisticians use a measure to determine whether two or more variables are related and also to determine the strength of the relationship between or among the variables. This measure is called a correlation coefficient. For example, there are many variables that contribute to heart disease, such as lack of exercise, smoking, heredity, age, stress, and diet. Of these variables, some are more important than others; therefore, if a nutritionist is interested in helping a patient, the nutritionist must know which factors are most important.

Another consideration in determining relationships is to ascertain what type of relationship exists. There are two types of relationships: simple and multiple.

In a simple relationship, there are only two variables under study. For example, a researcher may wish to know whether an increase in calcium intake actually increases bone density. This type of study involves a simple relationship, since there are only two variables, calcium intake and bone density.

In multiple relationships, many variables are under study. For example, an estimate of an adult patient's energy needs for basal energy expenditure may be related to body weight and height, and age. This type of study involves three variables.

Types of Relationship Between Two Variables

Simple relationships can be positive or negative. A positive relationship exists when both variables increase or decrease at the same time (Figure 7.1). For instance, a person's height and weight are related; and the relationship is positive, since the taller a person is, generally, the more the person weighs. In a negative relationship, as one variable increases, the other variable decreases, and vice versa (Figure 7.2). For example, if one compares the strength of people over 60 years of age, one will find that as age increases, strength generally decreases. There is no relationship between two variables (Figure 7.3).

Figure 7.1 : Positive relationship   Figure 7.2 : Negative relationship

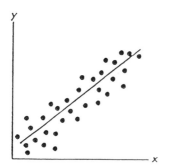
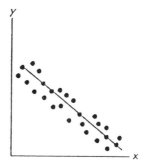

Figure 7.3 : No relationship

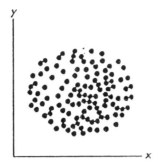

Applications of Correlation and regression in Research

Figure 7.4 shows the dose-response relationship between physical activity and the risk of colon cancer from a single large case-control study. This shows the steady decline in risk with higher lifetime levels of activity.

Figure 7.4

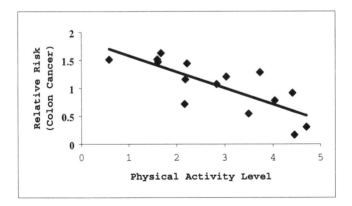

Epidemiological evidence for a relationship between bone fractures and usual level of dietary protein is consistent with the adverse effect of protein on calcium balance. Figure 7.5 shows a positive relationship between hip fracture incidence in women over age 50 years of age and estimated per capita dietary protein intake.

Figure 7.5

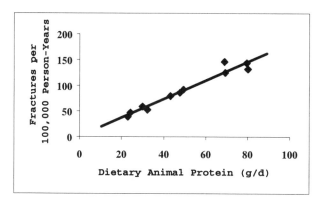

Examination of the change in total body calcium after the menopause shows a steady decrease in the loss of bone. Figure 7.6 shows a negative relationship between total body calcium and years of post-menopausal.

Figure 7.6

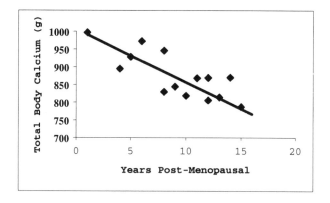

To determine the strength of the relationship between two variables, researchers use the correlation coefficient (*r*), which we will discuss in the following section.

Finally, the fourth question asks what type of prediction can be made. Predictions are made in all areas and on a daily basis. Examples include disease prediction, i.e. incidence of cancer, diabetes, osteoporosis, and heart disease. Some predictions are more accurate than others, due to the strength of the relationship. That is, the stronger the relationship is between variables, the more accurate the prediction is (regression).

How Correlation Research Investigates

As the above examples, in simple correlation and regression studies, the researcher collects data on two variables to see whether a relationship exists between the variables. For example, if a researcher wishes to see whether there is a relationship between the amount of saturated fat intake and plasma cholesterol levels. We assume the researcher collected data as follows:

Example 7.1

Subject	Saturated Fat Intake (x) g	Cholesterol Level (y) mg/dl
A	10	135
B	13	145
C	20	210
D	18	150
E	17	150

The two variables for this study are called the independent variable and the dependent variable. The independent variable is the variable in regression that can be controlled or manipulated. In this case, the variable, **saturated fat intake**, is the independent variable and is designated as the x variable. The dependent variable is the variable in regression that cannot be controlled or manipulated. **Plasma cholesterol** level is the dependent variable, designated as the y variable. The reason for this distinction between the variables is that one assumes that plasma cholesterol levels depends on the amounts of saturated fat intake. Also, one assumes that, to some extent, the people can regulate or control the amounts of fat intake.

The determination of the x and y variables is not always clear-cut and sometimes is an arbitrary decision. For example, if a researcher studies the effects of age on a person's blood pressure, the researcher can generally assume that a person's age affects the person's blood pressure. Hence, the variable "age" can be called the independent variable and the variable "blood pressure" can be called the dependent variable.

The independent and dependent variables can be plotted on a graph called a scatter plot. A scatter plot is an aid for understanding the correlation and regression techniques. A scatter plot is a graph of the independent and dependent variables in regression and correlation analysis. The scatter plot for example 7.1 is shown in Figure 7.7.

Exercise 7.1

Construct a scatter plot for the data obtained in a study of age and systolic blood pressure of six randomly selected subjects. The data are shown in the following table. Draw and label the x and y axes. What relationship exists between the two variables?

Subject	Age, x (yrs)	Pressure, y (mmHg)
A	43	128
B	48	120
C	56	135
D	61	143
E	67	141
F	70	152

Figure 7.7

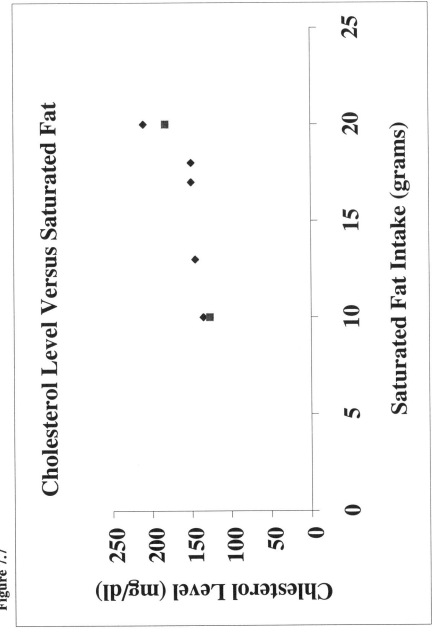

Cholesterol Level Versus Saturated Fat

Exercise 7.2

Construct a scatter plot for the data obtained in a study of age and femur bone density of seven randomly selected subjects. The data are shown in the following table. Draw and label the x and y axes. What relationship exists between the two variables?

Subject	Age, x	Bone density, y (g/cm^2)
A	30	1.02
B	35	0.95
C	47	0.91
D	50	0.88
E	60	0.79
F	70	0.75
G	80	

Exercise 7.3

Construct a scatter plot for the data obtained in a study on the number of hours a person exercises each week and the amount of milk (oz) each person consumes per week. The data follow. Draw and label the x and y axes. What relationship exists between the two variables?

Subject	Hours, x	Amount, y
A	3	48
B	0	8
C	2	32
D	5	64
E	8	10
F	5	32
G	10	56
H	2	72
I	1	48

Correlation Coefficient

Statisticians use a measure called the correlation coefficient to determine the strength of the relationship between two variables. There are several different types of correlation coefficient. The one explained in this section is called the Pearson product moment correlation coefficient.

The correlation coefficient computed from the sample data measures the strength and direction of a relationship between two variables. The symbol for the sample correlation coefficient is r. The symbol for the population correlation coefficient is ρ.

The range of the correlation coefficient is from -1 to +1. If there is a strong, positive linear relationship between the variables, the value of r will be close to +1. If there is a strong, negative linear relationship between the variables, the value of r will be close to -1. When there is no linear relationship between the variables or only a weak relationship, the value of r will be close to 0.

The formula is

$$r = \frac{N\Sigma XY - (\Sigma X)(\Sigma Y)}{\sqrt{[N\Sigma X^2 - (\Sigma X)^2][N\Sigma Y^2 - (\Sigma Y)^2]}}$$

where N is the number of paired scores.

Example 7.2

Compute the value of the correlation coefficient for the data obtained in the study of age and blood pressure given in Exercise 7.1

Subject	Age, X	Pressure, Y	XY	X^2	Y^2
A	43	128	5,504	1,849	16,384
B	48	120	5,760	2,304	14,400
C	56	135	7,560	3,136	18,225
D	61	143	8,723	3,721	20,449
E	67	141	9,447	4,489	19,881
F	70	152	10,640	4,900	23,104
	ΣX=345	ΣY=819	ΣXY=47,634	ΣX^2=20,399	ΣY^2=112,443

Substitute in the formula and solve for r

$$r = \frac{N\Sigma XY - (\Sigma X)(\Sigma Y)}{\sqrt{[N\Sigma X^2 - (\Sigma X)^2][N\Sigma Y^2 - (\Sigma Y)^2]}}$$

$$= \frac{(6)(47,634) - (345)(819)}{\sqrt{[(6)(20,399) - (345)^2][(6)(112,443) - (819)^2]}} = 0.897$$

The correlation coefficient suggests a strong positive relationship between age and blood pressure.

Exercise 7.4

Compute correlation coefficients for exercises 7.1, 7.2, and 7.3 compare those coefficients.

What the Coefficient of Correlation Means

Interpreting Reliability of r

First, there are several ways of interpreting r. One criterion is its reliability, or significance. Does it represent a real relationship? That is, if the study were repeated, what is the probability of getting a similar relationship? For this statistical criterion of significance, simply consult a table. In using the table, select the desired level of significance, such as the 0.05 level, and then enter the table in accordance with the appropriate degrees of freedom (df) (df are based on the number of subjects corrected for bias), which, for r, is equal to N-2. Table A.3 in Appendix A contains the necessary correlation coefficients for significance at the 0.05 and 0.01 levels. Refer to the example of the correlation above (r = 0.897). The degrees of freedom are N-2 = 6-2 = 4 (remember, the variable N in correlation refers to the number of pairs of scores). When entering the table at 4 df, we see that a correlation of 0.8114 is necessary for significance of a two-tailed test at the 0.05 level (and 0.9172 at the 0.01 level). Therefore, we would have to conclude that our correlation of 0.897 is significant at the 0.05 level, but not significant at the 0.01 level.

Another glance at Table A.3 reveals a couple of obvious facts. The correlation needs for significance decreases with increased numbers of subjects (df). In our example, we had only 6 subjects (or pairs of scores). However, if four more subjects had been in the sample (N=10), then there would be 8df, and the correlation required for significance at the 0.05 level for 8df would be 0.6319, and at the level 0.01 would be 0.7646. Our correlation of 0.897 meets that test of significance at the 0.01 level. Notice, however, that very low correlation coefficients can be significant if you have large sample of subjects. At the 0.05 level, a correlation of 0.38 is significant with 25df, r=0.27 is significant with 50 df, and 0.195 is significant with 100 df. In fact, with 1,000 df, a correlation of 0.08 is significant at the 0.01 level.

Example 7.3

Compute the value of the correlation coefficient for the data given in Exercise 7.2 for the age and femur bone density of seven subjects.

Subject	Age, X	Bone Density, Y (mg/cm^2)	XY	X^2	Y^2
A	30	10.2	306	900	104.04
B	35	9.5	332.5	1,225	90.25
C	47	9.1	427.7	2,209	82.81
D	50	8.8	440	2,500	77.44
E	60	7.9	474	3,600	62.41
F	70	7.5	525	4,900	56.25
G	80	7.1	568	6,400	50.41
	ΣX= 372	ΣY= 60.1	ΣXY= 3,073.2	ΣX^2= 21,734	ΣY^2= 523.61

$$r = \frac{N\Sigma XY - (\Sigma X)(\Sigma Y)}{\sqrt{[N\Sigma X^2 - (\Sigma X)^2][N\Sigma Y^2 - (\Sigma Y)^2]}}$$

$$= \frac{(7)(3,073.2) - (372)(60.1)}{\sqrt{[(7)(21,734) - (372)^2][(7)(523.61) - (60.1)^2]}}$$

$$= \frac{21,512.4 - 22,357.2}{\sqrt{[152,138 - 138,384] \times [3,665.27 - 3,612.01]}}$$

$$= \cfrac{-844.8}{\sqrt{[13,754][53,26]}} = \cfrac{-844.8}{855.88} = -0.987$$

The value of r suggests a strong negative relationship between age and femoral bone density. df = 7-2 = 5   Since two-tail test the significant correlation coefficient at the level of 0.05 is 0.7545; and at the level of 0.01 is 0.8745, therefore it can be concluded that femural bone density significantly decreased with aging (p<0.01).

Example 7.4

Compute the value of the correlation coefficient for the data given in Exercise 7.3 for the number of hours a person exercises and the number of milk a person consumes per week.

Subject	Hours, X	Amount, Y	XY	X^2	Y^2
A	3	48	144	9	2,304
B	0	8	0	0	64
C	2	32	64	4	1,024
D	5	64	320	25	4,096
E	8	10	80	64	100
F	5	32	160	25	1,024
G	10	56	560	100	3,136
H	2	72	144	4	5,184
I	1	48	48	1	2,304
	$\Sigma X= 36$	$\Sigma Y= 370$	$\Sigma XY= 1,520$	$\Sigma X^2= 232$	$\Sigma Y^2 = 19,236$

$$r = \cfrac{N\Sigma XY - (\Sigma X)(\Sigma Y)}{\sqrt{[N\Sigma X^2 - (\Sigma X)^2][N\Sigma Y^2 - (\Sigma Y)^2]}}$$

$$= \cfrac{(9)(1,520) - (36)(370)}{\sqrt{[(9)(232) - (36)^2][(9)(19,236) - (370)^2]}} = 0.067$$

The value of r indicates a very weak positive relationship between the variables. df = 9 - 2 = 7. The coefficient for 7 df at 0.05 is 0.6664. Therefore, there is no significant relationship between the number of hours a person exercises and the number of milk a person consumes per week.

Using Correlation for Prediction (Regression)

In studying relationships between two variables, the researcher, after collecting the data, constructs a scatter plot. The purpose of the scatter plot is to determine the nature of the relationship. The possibilities include a positive linear relationship, a negative linear relationship. A curvilinear, or no discernible relationship. After the scatter plot is drawn, the next step is to compute the value of the correlation coefficient and to test the significance of the relationship. If the value of the correlation coefficient is significant, the next step is to determine the equation of the **regression line,** which is the line of best fit of the data. The purpose of the regression line is to enable the researcher to see the trend and make **prediction** on the basis of the data.

Verducci (1980) provided one of the best example in introducing the regression equation concerning monthly salary and annual income. If there are no other sources of income, we can predict with complete accuracy the annual income of workers simply by multiplying their monthly salaries by 12. Figure 7.8 illustrates this perfect relationship. By plotting the monthly salary (the X, or predictor variable), the predicted annual income (the Y, or criterion, variable) can be obtained.

The formula can be $Y = 12X$

If all workers got an annual supplement of $500, the formula can be $Y = 500 + 12X$
This formula is the general formula for a straight line and is expressed as follows:

$Y = a + bX$

where Y = the predicted score, or criterion; a = the intercept; b = the slope of the

Figure 7.8

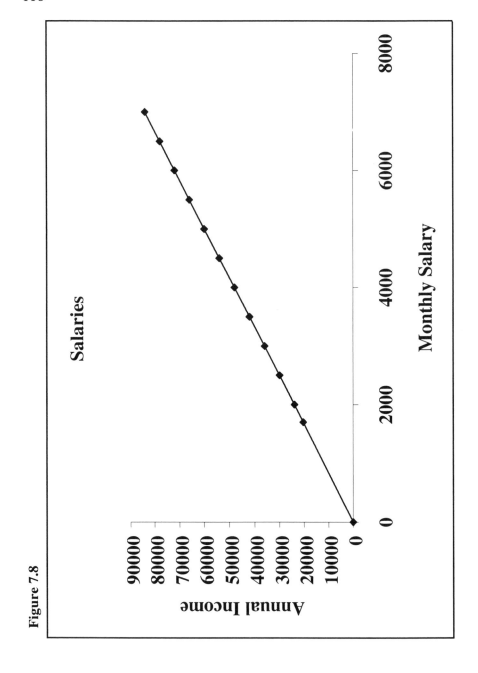

Salaries

regression line; and X = the predictor.

There are several methods for finding the equation of the regression line. We will use the following two formulas in this section. These formulas use the same values that are used in computing the values of the correlation coefficient.

$$a = \frac{(\Sigma Y)(\Sigma X^2) - (\Sigma X)(\Sigma XY)}{N(\Sigma X^2) - (\Sigma X)^2}$$

$$b = \frac{N(\Sigma XY) - (\Sigma X)(\Sigma Y)}{N(\Sigma X^2) - (\Sigma X)^2}$$

where a is the Y intercept and b is the slope of the line.

Example 7.5

Find the equation of the regression line for the data in Example 7.1, and graph the line on the scatter plot of the data.

The values needed for the equation are n = 6, ΣX =345, ΣY = 819, ΣXY = 47,634, and ΣX^2 = 20,399. Substituting in the formulas,

$$a = \frac{(819)(20,399) - (345)(47,634)}{(6)(20,399) - (345)^2} = 81.048$$

$$b = \frac{(6)(47,634) - (345)(819)}{(6)(20,399) - (345)^2} = 0.964$$

Y = 81.048 + 0.964X

Therefore,

The scatter gram and regression line for the above (Example 7.2) are shown in Figure 7.9.

Exercise 7.5

Find the equation of the regression line for the data in Exercises 7.1 & 7.2, and graph the lines on the scatter plots, respectively.

Interpreting Meaningfulness of r

The most commonly used criterion for interpreting the correlation coefficient as to meaningfulness is the coefficient of determination (r^2). In this method, the portion of common of the factors that influence the two variable is determined. The coefficient of determination is the ratio of the explained variation to the total variation and is denoted by r^2. That is

$$r^2 = \frac{\text{Explained variation}}{\text{Total variation}}$$

If $r = 0.8$, $0.8^2 = 0.64$, or 64%: then 64% is common (explained) variance, and 36% is error (unexplained) variance. 36% $[(1.00 - r^2) \times 100]$ is called the coefficient of nondetermination.

The coefficient of determination is a measure of the variation of the dependent variable that is explained by the line and the independent variable.

Standard Error of Estimate

When a Y' value is predicted for a specific X value, the prediction is a point prediction. However, a prediction interval about the Y' value can be constructed, just as a confidence interval was constructed for an estimate of the population mean. The prediction interval uses a statistic called the standard error of estimate.

The standard error of estimate, denoted by $s_{y.x}$ is the standard deviation of the observed Y values about the predicted Y' values. The formula is:

$$s_{y.x} = \sqrt{\frac{\Sigma (Y - Y')^2}{N - 2}} \qquad \text{or} \quad s_{y.x} = s_y \sqrt{1 - r^2}$$

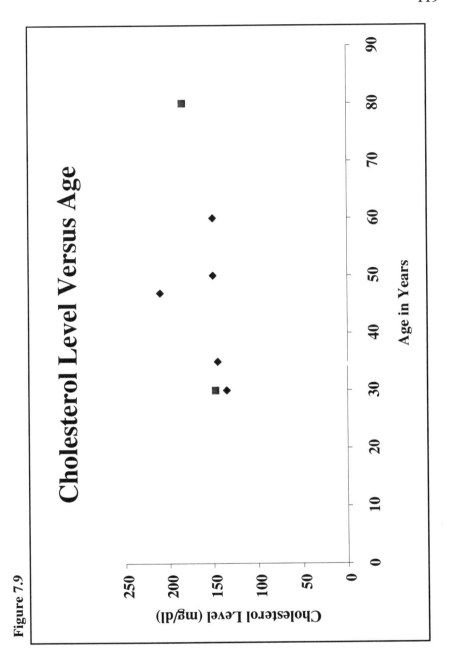

Figure 7.9

The standard error of estimate is similar to the standard deviation, but the mean is not used. As can be seen from the formula, the standard error of estimate is the square root of the unexplained variation - i.e. the variation due to the difference of the observed value and the expected values - divided by N-2. So the closer the observed values are to the predicted values, the smaller the standard error of estimate will be. The N-2 in the denominator is used in this case because our sample data consist of two variables.

Example 7.6

A pediatrician measured waist circumferences for the children aged 3 to 10 years, and obtained the following data. The regression equation was $Y' = 1.0277 + 5.1389\ X$. Compute the standard error of estimate.

Subject	Age, X	Waist, Y (cm)	Y'	$(Y - Y')$	$(Y - Y')^2$
A	6	30	31.86	-1.86	3.4596
B	9	49	47.28	1.72	2.9584
C	3	18	16.44	1.56	2.4336
D	8	42	42.14	-0.14	0.0196
E	7	39	37.00	2.00	4.0000
F	5	25	26.72	-1.72	2.9584
G	8	41	42.14	-1.14	1.2996
H	10	52	52.42	-0.42	0.1764
Total	56	296*	296.00*	0.00	17.3056

*Note that the sum of Y and Y' are equal. This must be true if $\Sigma(Y - Y') = 0$

The standard error of estimate is calculate as follows:

$$S_{y.x} = \sqrt{\frac{\Sigma(Y - Y')^2}{N - 2}} = \sqrt{\frac{17.3056}{6}} = \frac{17.3056}{2.449} = 7.07\ (cm)$$

The value of $s_{y.x}$ will always be expressed in the units of Y variable.

For instance, we predict that a 8 years child will score 42.14 ± 7.07 (round up of

7.07). To express it another way, the prediction range will be between 35.07 and 49.21 cm.

The larger the correlation, the smaller the error of prediction. Also the smaller the standard deviation of the criterion, the smaller the error.

Partial Correlation

The correlation between two variables is sometimes misleading and may be difficult to interpret when there is little or no correlation between the variables other than that caused by their common dependence on a third variable. For example, many attributes increase regularly with age from 6 to 18 years, such as height, weight, strength, mental performance, vocabulary, reading skills, and so on. Over a wide age range, the correlation between any two of these measures will almost certainly be positive and will probably be high because of the common maturity factor with which they are highly correlated. In fact, the correlation may drop to zero if the variability caused by age difference is eliminated. We can control this factor of age in one of two ways. We can select only children of the same age, or we can partial out the effects of age statistically by holding it constant.

For example, three variables are 1 = math achievement, 2 = strength, and 3 = age. Then, $r_{12.3}$ is the partial correlation between variables 1 and 2 with 3 held constant. We can make up some correlation coefficients between the three variables: $r_{12} = .80$; $r_{13} = .90$; and $r_{23} = .88$.

The formula for:

$$r_{12.3} = \frac{r_{12} - r_{13} r_{23}}{\sqrt{(1 - r_{13}^2)} \sqrt{(1 - r_{23}^2)}} = \frac{.80 - .90 \times .88}{\sqrt{(1 - .90^2)} \sqrt{(1 - .88^2)}} = 0.038$$

Therefore, there is no relationship between math achievement score and strength when age is partial out. Partial correlation is primarily used to develop a multiple regression equation with two or more predictor variables.

Multiple Regression Prediction Equation

The prediction equation resulting from multiple regression is basically that of the two-variable regression model, $Y' = a + bX$. The only difference is that there is more than one X variables; thus, the equation is

$$Y' = a + b_1 X_1 + b_2 X_2 + b_3 X_3 \dots \dots b_i X_i$$

An example of a multiple prediction formula follows. In this equation, a man's lean body weight (LBW) is being predicted from several anthropometric measures, including skinfolds thicknesses, circumferences, and diameters. The following formula, developed by Behnke and Wilmore (1974), has a correlation of .985 and a standard error of estimate of 2.358, which is interpreted just the same as in the regression equation with only one predictor variable.

LBW = 10.138 +0.9259(wt) - 0.1881(thigh skinfolds) + 0.637(bi-iliac diameter) + 0.4888(neck circumference) - 0.5951(abdominal circumference)

Example 7.7

A researcher collects the following data and determines that there is a significant relationship between ages of children and sum of neck and abdominal circumferences. The regression equation is $Y' = 55.57 + 8.13X$. Find the standard error of estimate.

Age, X	Neck circumference, Y	Y'	(Y - Y')	$(Y - Y')^2$
1	62	63.7	- 1.7	2.89
2	78	71.83	6.17	38.0689
3	70	79.96	- 9.96	99.2016
4	90	88.09	1.91	3.6481
5	93	88.09	4.91	24.1081
6	103	104.35	- 1.35	1.8225
Σ				169.7392

$$s_{y.x} = \sqrt{\frac{\Sigma(Y - Y)^2}{N - 2}} = \sqrt{\frac{169.7392}{6 - 2}} = 6.51$$

The standard deviation of observed values about the predicted values is 6.51

Exercise 7.6

The following data were obtained from a survey of the number of years a person smoked and the percentage of lung damage. What is correlation coefficient? Predict the percentage of lung damage for a person who has smoked for 30 years?

Years, X	Damage,%, Y	XY	X^2	Y^2
22	20			
14	14			
31	54			
36	63			
9	17			
41	70			
19	23			

Summary

We have explored some statistical techniques such as correlation and regression to determine relationships among variables. We introduced linear regressions, which can be used to predict one variable from another. Positive relationship, negative relationship, and no relationship among variables were discussed. We demonstrated how to apply these statistical techniques for predicting diseases based on dietary intake, e.g., sugar intake and colon cancer (positive relationship); physical activity and colon cancer (negative relationship).

The correlation coefficient is used to determine the strength of the relationship between two variables. We discussed the Pearson product moment correlation coefficient. The most commonly used criterion for interpreting the correlation coefficient as to meaningfulness is the coefficient of determination (r^2). The coefficient of determination is a measure of the variation of the dependent variable that is explained by the regression line and the independent variable.

Standard error of estimate is similar to the standard deviation, but the mean is not used. The standard error of estimate is the square root of the unexplained variation.

In multiple regression two or more predictor (independent variables) are used to predict the criterion (dependent) variable.

References

Abelow B.J., Holford T.R., Insogna, K.L. Cross-cultural association between dietary animal protein and hip fracture: a hypothesis. Calcif Tissue Intl 1992; 50: 14-20

Bartz, A. E. Basic Statistical Concepts. (4th ed.) New York: Macmillan, 1998.

Behnke, A.R., Wilmore, J.H. Evaluation and Regulation of Body Build and Composition. Englewood Cliffs, NJ: Prentis-Hall, 1974.

Benito E., Obrador A., Stigelbout A. A population-based case-control study of colorectal cancer in Majorca: I. Dietary factors. Intl J Cancer 1990; 45: 69-76

Cassidy A., Bingham S.A., Cummings S.H. Starch intake and colorectal cancer – an international comparison. British J Cancer 1994; 69: 937-942

Cohen, J., Cohen, P. Applied Multiple Regression in Behavioral Research. New York: Holt, Rinehart & Winston, 1983.

Gad, S.C. Statistics and Experiment Design for Toxicologists. (3rd ed.) Boca Raton, London, New York, Washington, D.C.: CRC press, 1999.

Gallagher J.C., Goldgar D., Moy A. Total bone calcium in normal women: effect of age and menopause status. J Bone Miner Res 1987; 2: 491-496

Mattson, D.E. Statistics: Difficult Concepts, Understandable Explanations. St. Louis, MO: C.V. Mosby, 1981.

Pedhazur, E.J. Multiple Regression in Behavioral Research. New York: Holt, Rinehart & Winston, 1982.

Potter, J.D. Chair of the Panel. Food, Nutrition and the Prevention of Cancer: a Global Perspective. Washington, D.C.: World Cancer Research Fund/American Institute for Cancer Research, 1997.

Pyczak, F. Statistics with a Sense of Humor. A Humorous Workbook and Guide to Study Skills. Los Angeles, CA: Pyrczak Publishing, 1989.

Sanders, D.H. Statistics: A First Course. (5th ed.) New York: McGraw-Hill, Inc. 1995.

Slattery M.L., Edwards S.L., Ma K-N, Friedman G.D., Potter J.D. Physical activity and colon cancer: a public health perspective. Ann Epidemiol 1997; 7: 137-145

Verducci, F.M. Measurements Concepts in Physical Education. St. Louis, MO:C.V. Mosby, 1980.

CHAPTER 8

===

DIFFERENCES AMONG GROUPS

Introduction

Statistical techniques are used for describing and finding relationships among variables, as we discussed in Chapters 6 and 7. They are also used to detect differences among groups. The latter are most frequently used for data analysis in experimental and quasi-experimental research. They enable us to evaluate the effects of an independent [cause or treatment or categorical variables (gender, age, race, etc.)] variable on a dependent variable (effect, outcome).

How Statistics Test Differences

In experimental research, the experimenter may establish the levels of the independent variable. For example, the experiment might involve the investigation of the effects of intensity of training on bone density. Thus, intensity of training is the independent variable (or treatment factor), whereas a measure of bone density is the dependent variable. Intensity of training could have any number of levels. If it were evaluated as a percentage of the 1 repetition maximum (1 RM), then it could be 30%, 40%, 50%, and so forth. The investigator would choose the number and the intensity of levels. In a simple experiment, the independent variable was two levels of intensity of training, for example, high intensity weight training (8 repetitions at 80% of the 1 RM), and low intensity weight training (16 repetitions at 40% of the 1 RM). The dependent variable was changes in bone density (Fetters).

The purpose of the statistical test is to determine whether the null hypothesis can be reject subject to the condition that the probability of making an error in rejecting the null hypothesis is less than a predetermined value (e.g., $p < 0.05$). In other words, do the two levels of treatment differ significantly ($p < 0.05$) so that these differences would not be attributable to a chance occurrence more than 5 times in 100? The statistical test is always of the null hypothesis. All that statistics can do is reject or fail to reject the null hypothesis. Statistics cannot accept the statistical hypothesis. Only logical reasoning good experimental design and appropriate theorizing can do so. Statistics can determine only whether the groups are different, not why they are different.

When you use statistics that test differences among groups, you want to establish not only whether the groups are significantly different but also the strength of the association between the independent and dependent variables, or the size of the difference between two groups. The t and F ratios are used throughout this chapter to determine whether groups are significantly different. Omega squared (ω^2) is used to estimate the degree of association between the independent and dependent variables. To some extent, ω^2 is similar to r^2 for the correlation presented in Chapter 7; both represent the same idea, which is present variance accounted for.

Three Types of Student's t Tests

The Student's t-distribution was developed around the turn of the 20th century by William Gusset, who worked for the Guinness Brewery, he was prevented by company policy from publishing articles under his real name and thus chose the pseudonym "Student." The t- distributions are appropriate when the distribution is unimodal, is symmetric, and has a standard deviation that is not excessive.

We discuss three types of t tests: One population t-test, t test for independent groups, and t-test for dependent groups.

One Population t-Test

First we want to consider the case in which the population mean is predicted based on some theoretical calculation or on some previous observations. For example, the graduate record examination (GRE) is constructed in such a way that the mean for a year of national administration is 500.

If 32 graduates of this department take the examination, we can think of them as a random sample from the population of graduates from this department. We can test the null hypothesis that the mean score on the GRE of the population of graduates of this department, μ, is equal to the national mean versus the alternate hypothesis that μ is different from the national mean. Symbolically we write this as

$H_o: \mu = 500$ versus $H_1 : \mu \neq 500$.

In order to write a general formula we use the general null and alternate hypotheses

$H_o: \mu = \mu_o$ versus $H_1 : \mu \neq \mu_o$

We need to keep in mind that μ_o is always some number (such as 500). We know that the sample mean M should be close to the true value of μ. Therefore we estimate $\mu - \mu_o$ by $M - M_o$ and calculate the t-test by the formula 8.1

$$t = \frac{M - \mu}{s_m / \sqrt{n}} \qquad \text{(Formula 8.1)}$$

Where s_m = the standard deviation from the sample and n = the number of subjects in the sample. Suppose that our 32 students took the GRE and earned a mean score of 600 with a standard deviation of 90. Then,

$$t = \frac{600 - 500}{90 / \sqrt{32}} = \frac{100}{15.9} = 6.29$$

Is the value of 6.29 significant? To find out you will need to check Table A.5 in Appendix A. To use the table, you need to know the degrees of freedom (df). Degrees of freedom are based on the number of subjects with a correction for bias: df = n - 1

In this case, df = 32 - 1= 31df is used to enter a t table to determine whether the calculated t is as large as or greater than the tabled t value. Note that across the top of Table A.5 are probability levels. We want to know if the calculated t value is significant at p<0.05 and p<0.01 levels. Read across to the 0.05 level for df = 30 since df = 31 is not in the Table. Now read down the left side (df) to the number in the t test from 31. However, 31df is not in the table. Therefore, read 2.042 where the 30 df row and the 0.05 column intersect. Is the calculated value (6.29) larger than this value (2.042)? Yes, it is. So t test is significant at p<0.05. How about at the level of 0.01? The calculated value 6.29 is also larger than the tabled value 2.75 at p<0.01. Thus, the nutrition class has a significantly higher average score on the nutrition test as compared to that of the population.

Independent t Test

The previous t test applied to determine whether population mean is different from a predicted value is not used very frequently. The most frequently used t test determines whether two sample means differ reliably each other. This is called an independent t test.

Using the Independent t Test

Suppose we return to our example at the beginning of this chapter: Do the two weight training intensities (80% and 40% of 1-RM) elicit a different adaptation in the bone mineral density of the lumbar spine, proximal femur, and total body when the total volume of work is constant? Let us further assume that there were 30 subjects who were randomly assigned to form the two groups of 15 each.

Formula 8.2 is the t-test for two independent samples:

$$t = \frac{M_1 - M_2}{\sqrt{s_1^2/n_1 + s_2^2/n_2}} \qquad \text{(Formula 8.2)}$$

Usually s_1 and s_2 are pooled using the formula

$$S_p = \sqrt{\frac{(n_1 - 1)\, s_1^2 + (n_2 - 1)s_2^2}{n_1 + n_2 - 2}}$$

Formula 8.3 is the version of the t-test formula most easily performed with a calculator:

$$t = \frac{M_1 - M_2}{\sqrt{\dfrac{[\Sigma X_1^2 - (\Sigma X_1)^2/n_1 + \Sigma X_2^2 - (\Sigma X_2)^2/n_2\,].\,(1/n_1 + 1/n_2)}{n_1 + n_2 - 2}}} \qquad \text{(Formula 8.3)}$$

The degrees of freedom for an independent t test are calculated as follows:

In this example, df = 15 + 15 - 2 = 28
$$df = n_1 + n_2 - 2$$

Training and Bone density Example:

Known Values: Group 1 (40%):   Group 2 (80%)
Mean density: M1 = 1.154 (g/cm^2):   M2 = 1.256 (g/cm^2)
Standard deviation: $s_1 = .05$: $s_2 = .04$
Number of subjects: n1 = 15: $n_2 = 15$

$$t = \frac{M_1 - M_2}{\sqrt{s_1^2/n_1 + s_2^2/n_2}} = \frac{1.154 - 1.256}{\sqrt{(.05)^2/15 + (.04)^2/15}} = \frac{0.102}{\sqrt{.0001666 + .0001066}}$$

$$= \frac{.102}{.01653} = 60.496$$

t(28) = 60.496, p<0.01

Thus, you can see that the 80% intensity of training allowed subjects to reach reliably higher bone density (M=1.256 g/cm^2) than did the 40% intensity training (M=1.154 g/cm^2).

Estimating Meaningfulness of Treatments

How meaningful is this effect? Or, stated more simply, is the increase in bone density of an addition .102 g/cm^2 worth the additional work of training 80% of 1 RM as compared to 40% of 1 RM? Given the total variation in changing bone density of the two groups, what we really want to know is how much of this variation is accounted for by (associated with) the difference in the two levels of the independent variable (80% vs. 40%).

Omega Squared (ω^2). One way to estimate this variation is to use the following formula (Tolson, 1980) to calculate ω^2:

$$\omega^2 = \frac{t^2 - 1}{t^2 + n_1 + n_2 - 1} \qquad \text{(Formula 8.4)}$$

We can apply the above example to this formula:

Known values
Differences between groups t = 60.496
Number of subjects in Group 1: 15
Number of subjects in Group 2: 15

$$\omega^2 = \frac{t^2 - 1}{t^2 + n_1 + n_2 - 1} = \frac{3659.766 - 1}{3659.766 + 15 + 15 - 1} = \frac{3658.766}{3688.766} = 0.9918$$

We can conclude that ω^2 = .9918 means 99.18% of the total variance in the bone density change can be accounted for by the difference in the two group's levels of training. The remaining variance, 0.82%, is accounted by other factors.

Dependent t Test

We have now considered use of the t test to evaluate whether a sample differs from a population and whether two independent samples differ from each other. A third application is called a **dependent t test.** This means that the two groups of scores are related in some manner. Usually, the relationship takes one of two forms:

1. two groups of subjects are matched on one or more characteristics and thus are no longer independent, or
2. one group of subjects is tested twice on the same variable, and the experimenter is interested in the change between the two tests.
3. subjects in one group are selected because they have a prescribed relationship with an individual in other group. For example, we might study mother – daughter pairs.

$$\text{The formula } t = \frac{\Sigma\,D}{\sqrt{[N\Sigma D^2 - (\Sigma D)^2]/(N-1)}} \qquad \text{(Formula 8.5)}$$

where N = the number of paired observations, and D = the posttest minus the pretest for each subject. Ten elderly subjects were treated with a cholesterol lowering diet for 15 wks. The plasma cholesterol levels for pre- and post-test are presented in the Table 8.1. Interpret the experimental results by calculating the dependent t test.

Table 8.1

Subjects	Posttest score	Pretest score	Posttest- Pretest	
	mg/dl	mg/dl	D	D^2
1	150	155	- 5	25
2	213	223	-10	100
3	180	179	1	1
4	209	225	-16	256
5	180	175	5	25
6	256	270	-14	196
7	189	193	- 4	16
8	245	250	- 5	25
9	222	234	- 12	144
10	225	222	3	9
	2,069	2,126	57	797

Sum of posttest score Σ post = 2,069
Sum of pretest score Σ pre = 2,126
Sum of D: Σ D = - 57
Sum of D^2: $\Sigma D^2 = 797$
Number of paired observations: N =10

Posttest mean = 206.9
Pretest mean = 212.6

$$\text{The formula } t = \frac{\Sigma\,D}{\sqrt{[N\Sigma\,D^2 - (\Sigma\,D)^2]/(N-1)}} = \frac{57}{\sqrt{[10 \times 797 - (57)^2]/9}}$$

$$= \frac{57}{\sqrt{524.55}} = 2.489$$

The results indicate that the posttest mean (206.9 mg/dl) was significantly lower than the pretest mean (212.6 mg/dl), t(9) = 2.489, p<0.05. We can conclude that dietary treatment significantly decreased plasma cholesterol at p<0.05.

Exercise 8.1

In order to determine whether or not a particular heat treatment is effective in reducing the number of bacteria in skim milk, counts were made before and after treatment on 12 samples of skim milk with the following results. The data are presented as of direct microscopic counts. Interpret the effect of heat treatment on bacteria in skim milk. (Table 8.3)

Table 8.3

Sample	Before Treatment	After Treatment	Post - Pre Test (D)	D^2
1	698	695		
2	708	694		
3	834	817		
4	530	505		
5	626	629		
6	677	681		
7	703	649		
8	556	534		
9	597	598		
10	664	650		
11	703	683		
12	769	699		

Sum of posttest score	Σ post =
Sum of pretest score	Σ pre =
Sum of D:	$\Sigma D =$
Sum of D^2:	$\Sigma D^2 =$
Number of paired observations :	N =

Posttest mean =
Pretest mean =

$$t = \frac{\Sigma D}{\sqrt{[N\Sigma D^2 - (\Sigma D)^2]/(N-1)}} =$$

The formula

Explain the results:

t Tests and Power in Research

In chapter 6, power was mentioned as the probability of rejecting the null hypothesis when null hypothesis is false. To obtain power in research is very desirable, as the odds of rejecting a false null hypothesis are increased. The independent t test is used here to explain three ways to obtain power (in addition to setting the alpha level). However, these ways apply to all types of experimental research.

Consider the formula for the independent t test:

$$t = \frac{M_1 - M_2}{\sqrt{\dfrac{s_1^2}{n_1} + \dfrac{s_2^2}{n_2}}}$$

.... 1

.... 2

.... 3

Note that we have placed (1,2,3) beside the three horizontal levels of this formula. These three levels represent what can be manipulated to increase or decrease power.

1.   The first level ($M_1 - M_2$) gives power if we can increase the difference between the two means. How can the difference between the means increase? It could be by applying stronger more concentrated treatments.

2.   The second level is s_1^2, s_2^2, or the variance for each of the two groups. Recall that the standard deviation represents the spread of the scores about the means. If this spread becomes smaller, the variance is also smaller. How can the standard deviation and thereby the variance be made smaller? The answer is to apply the treatments more consistently.

3.   Finally, the third level (n_1, n_2) is the number of subjects in each group. If n_1, and n_2 are increased and the first and second levels remain the same, the denominator will become smaller and the t ratio will become larger, thus increasing the odds of rejecting the null hypothesis and obtaining power.

Power may be obtained by using strong treatments, administering those treatments consistently, using as many subjects as feasible.

    The t ratio has a numerator and a denominator. From a theoretical point of view, the numerator is regarded as true variance, or the real differences between the means. The denominator is considered error variance, or variation about the mean. Thus, the t ratio is

        True variance
t = ----------------------   (Formula 8.6)
        Error variance

where true variance = $M_1 - M_2$ and error variance = $\sqrt{s_1^2/n_1 + s_2^2/n_2}$.
When a significant t ratio is found, we are really saying that true variance exceeds error variance to a certain degree.

The estimate the strength of the relationship (ω^2) between the independent and dependent variables is represented by the ratio of true variance to total variance.

         True variance
ω^2 =------------------------   (Formula 8.7)
         Total variance

Thus, ω^2 represents the proportion of the total variance that is due to the treatments (true variance).

Analysis of Variance (ANOVA)

    As we discussed previously, t tests were used when the means of two populations were compared; however, many statistical studies involve comparing two variances or standard deviations. For example, a researcher may be interested in comparing the variances of the cholesterol of men with the variance of the cholesterol of women. For the comparison of two variance or standard deviation, an F test is used. The F test should not be confused with the chi-square test, which compares a single sample variance to a specific population variance, which we will discuss in Chapter 9.

Another use for the F test is in a statistical technique called analysis of variance (ANOVA). This technique is used to test hypotheses involving three or more means (Note: t test is used for comparing one or two means). For example, suppose a nutritionist wishes to know whether three different dietary fat levels differently affect plasma cholesterol. The nutritionist will use the ANOVA technique for this test.

For three groups, the F test can only indicate whether or not a difference exists among the three means. It cannot indicate where the difference lies - i.e., which group is different from which group. If F test indicates that there is a difference among the means, other statistical tests are used to find out where the difference exists. The most commonly used tests are the Scheffe's, Duncan's, Newman-Keuls', and Tukey's multiple range tests.

χ) The ANOVA that is used to compare three or more means called a one-way ANOVA since it contains only one variable. In the above example, the only variable is the level of fat in the diets. The ANOVA can be extended to studies involving two or more variables. For instance, the effects of the type of dietary fiber - oat bran and wheat bran, and type of fat - beef tallow and fish oil on plasma cholesterol can be studied. In this study has two variables such as type of fat and type of fiber, and it is a two-way analysis of variance (two independent variables).

If two independent samples are selected from normally distributed populations in which the variances are equal ($\sigma_1^2 = \sigma_2^2$) and if the sample variances s_1^2 and s_2^2 are compared as s_1^2/s_2^2, the sampling distribution of the variance is called F distribution.

Characteristic of the F Distribution:

1. The F values cannot be negative, because variances are always positive or zero.
2. The distribution is positively skewed.
3. The mean value of F is approximately equal to 1.
4. The F distribution is a family of curves based upon the degrees of freedom of the variance of the numerator and the degrees of freedom of the variance of the denominator.

Calculating Simple ANOVA

Table 8.4 provides the formula for calculating simple ANOVA and the F ratio. This method, so called ABC method, is simple.

$A = \Sigma X^2$ ---sum of square of each score

$B = (\Sigma X)^2/N$ ---sum each score, square the sum, and divide by the total number of subjects

$C = (\Sigma X_1)^2 /n_1 + (\Sigma X_2)^2 /n_2 + ... (\Sigma X_i)^2 /n_i$ --- sum each subject score in group 1, square the sum, divide by the number of subjects in group 1; do the same for group 2, and so on; then sum all the group sums

Next fill in the summary table for ANOVA using A, B, and C. Thus, the between groups (true variance) sum of squares (SS) is equal to C - B; the between-groups degrees of freedom (df) is the number of groups minus one (k-1); the between groups variance or mean square is (C - B)/k-1. The within groups (error variance) SS is equal to A - C. The degrees of freedom are N-k. Thus, the within groups variance or mean square is (A - C)/N-k. The same follows for the total. The F ratio is MS_b/MS_w.

Table 8.4

$A = \Sigma X^2$
$B = (\Sigma X)^2/N$
$C = (\Sigma X_1)^2 /n_1 + (\Sigma X_2)^2 /n_2 + ... (\Sigma X_i)^2 /n_i$

Summary table for ANOVA

Source	SS	Df	MS	F
Between (true)	C - B	k-1	$MS_b=(C-B)/(k-1)$	MS_b/MS_w
Within (error)	A - C	N-k	$MS_w= (A-C)/(N-k)$	
Total	A - B	N-1		

where X=a subject score, N=total number of subjects, n= number of subjects in a group, k=number of groups, SS=sum of square, df= degrees of freedom, MS= mean square

The F ratio is Ms_b/MS_w (true/error).

Example 8.1

The comparative scores to contribute energy intake by fat groups, milk groups and alcohol beverages for a meal are presented in the following table. Interpret whether those food groups differently contribute to energy source by using ANOVA.

Subject	Fat		Milk		Alcohol	
	X	X^2	X	X^2	X	X^2
1	4.3	18.49	2.2	4.84	0.2	0.04
2	4.2	17.64	2.1	4.41	0.1	0.01
3	4.9	24.01	1.7	2.89	0.1	0.01
4	5.2	27.04	1.8	3.24	0.2	0.04
5	3.8	14.44	2.5	6.25	0.3	0.09
6	2.9	8.41	1.8	3.24	0.1	0.01

$\Sigma X=25.3$ $\Sigma X^2=110.03$ $\Sigma X=12.1$ $\Sigma X^2=24.42$ $\Sigma X=1$ $\Sigma X^2=0.2$
M1=4.22   M2 = 2.01   M3 =0.167

$A = \Sigma X^2 = 110.03 + 24.42 + 0.2 = 134.65$

$B = (\Sigma X)^2/N = (25.3 + 12.1 + 1)^2 /18 = 1474.56/18 = 81.92$
$C = (\Sigma X_1)^2 /n_1 + (\Sigma X_2)^2 /n_2 + ... (\Sigma X_i)^2 /n_i$
$= (25.3)^2 /6 + (12.1)^2/6 + (1)^2 /6$
$= 640.09/6 + 146,41/6 + 1/6 = 106.68 + 24.40 + 0.167 = 131.247$

Summary Table for ANOVA

Source	SS	df	MS	F
True	131.247 - 81.92 =49.327	2	24.663	108.65
Error	134.65 - 131.247=3.403	15	0.227	
Total	134.65 - 81.92 = 52.73	17		

The F ratio is 108.65. Table A.6 in Appendix A contains tabled F values for the 0.05 and 0.01 levels of significance. Although the numbers in this table are obtained the same way as in the t table, you use the table in a slightly different way. Note in Table 8.4, that F ratio is obtained by dividing Ms_b by Ms_w. The term Ms_b has 2 df associated with it (numerator) and the Ms_w has 15 df associated with denominator. Notice also that the F table has degrees of freedom across the top (numerator) and down the left hand column (denominator). For our F of 108.65, read down the 2- df column to the 15-df row; there are two numbers. The top number (3.68) is the tabled F for the 0.05 level, whereas the bottom number (6.36) is the tabled F for the 0.01 level term. If our alpha had been established as 0.05, then you can see that our value 108.65 is larger than the tabled value of 0.05 (actually, it is also larger than 0.01). So our F is significant. Therefore, we know that there are significant differences among those means.

Follow-up Testing

We now know that significant differences exist among the three group means. However, we do not know whether all three groups differ. Thus, we next perform a follow-up test. One way to do this is two use t tests between groups 1 and 2, 1 and 3, and 2 and 3. Several follow-up tests protect the experiment-wise error rate. These methods include Scheffe, Newman-Keuls, Duncan's, and several others. Each of the tests is calculated in a slightly different way, but they all are conceptually similar to the t test in that they identify which groups differ from each other. The Scheffe method is the most conservative, followed by Turkey, which means it identifies fewer significant differences. Duncan's is the most liberal, identifying more significant differences. Newman-Keuls falls between the two.

We demonstrate the use of the Newman-Keuls method making multiple comparisons among means. Table 8.5 arranges the means from highest (4.22) to lowest (0.167). By following up steps 1-8, you can determine how they are different each other.

Step 1 Calculate error term $E= \sqrt{Ms_w/n} = \sqrt{0.227/6} = .1945$

Step 2 Order the means form highest to lowest: 4.22, 2.01, 0.167

Step 3 Calculate the differences between the means; 2.21, 4.053, 1.843

Step 4 Divide the difference by E and enter in table above:
   2.21/.1945=11.362;
   4.053/.1945=20.838;
   1.843/.1945=9.476

Step 5 Calculate the steps between means; this is the number of means in the ordered set --- fat group and milk is 2 steps, fat group and alcohol beverages is 3 steps, and milk group and alcohol beverages is 2 steps. The number of steps is called k.

Step 6 The df for error is 15.

Step 7 Enter the Studentized range table (Table A.7 in Appendix A) with k and 15 df depending on which groups are being compared --- fat and milk would be 2 and 15df, fat and alcohol would be 3 and 15 df, and milk an alcohol would be 2 and 15df.

Step 8 Compare value calculated for the table above with value in Table A.7 at alpha 0.01. Since 15 df is not in the table, we will get 4.13 for 2 steps 4.78 or 3 steps. We can see that11.362 and 9.476 are larger than 4.13 for 2 steps, and 20.838 is larger than 4.78 for 3 steps.

Table 8.5 Newman-Keuls Test

Group	M	2	3
Fat	4.22	11.362**	20.838**
milk	2.01	9.476**	
Alcohol	0.167		

**p < 0.01

Therefore, all three groups differ from one another at p<0.01 level. We could concluded that the fat group contribute significantly more energy intake than either milk group or alcohol beverages, and milk group contribute significantly more than alcohol beverages.

Two-Way ANOVA (Factorial Analysis)

The ANOVA technique shown previously is called a one-way ANOVA since there is only one independent variable. The two-way ANOVA is an extension of the one-way ANOVA; it involves two independent variables. The independent variables are also called factors.

The two-way ANOVA is quite complicated, and many aspects of the subject should be considered when one is using a research design involving a two-way ANOVA. In doing a study that involves a two-way ANOVA, the researcher is able to test the effects of two independent variables or factors on one dependent variable (**main effects). Main effects are tests of each independent variable when the other is disregarded (and controlled).**

In addition, the **interaction effect of the two variables can also be tested.** For example, suppose a researcher wishes to test the effects of two different types of dietary fibers and two different types of dietary fats on plasma cholesterol levels. The two independent variables are the type of dietary fiber and the type of dietary fats, while the dependent variable is the cholesterol level of rabbits. Other factors, such as other dietary components, age, sex, strains, etc., are held constant.

In order to conduct this experiment, the researcher sets up four groups: Assume that dietary fibers are designed by the letters A1 and A2 and dietary fats by B1 and B2. The groups for such a two-way ANOVA are sometimes called treatment groups. The four groups are as follows:

Group 1 dietary fiber A1, dietary fat B1
Group 2 dietary fiber A1, dietary fat B2
Group 3 dietary fiber A2, dietary fat B1
Group 4 dietary fiber A2, dietary fat B2

| | | Dietary fat | |
		B1	B2
Dietary fiber	A1	A1xB1	A1xB2
	A2	A2xB1	A2xB2

The animals are assigned to the groups at random. This design is called a 2 x 2 factorial design, since each variable consists of two levels, i.e. two different treatments.

The two-way ANOVA enables the researcher to test the effects of dietary fiber and fat in a single experiment rather than in separate experiments involving the type of dietary fiber alone and the type of dietary fat alone. Furthermore, the researcher can test an additional hypothesis about the effect of interaction of the two variables, type of fats and type of fiber, on cholesterol. For example, is there a difference in cholesterol levels using fiber A1 and fat B2 and the cholesterol levels using fiber A2 and fat B1? When a difference of this type occurs, the experiment is said to have a significant interaction effect. That is, the types of fat affect the cholesterol levels differently in different fiber types.

There are many different kinds of two-way ANOVA designs, depending on the number of levels of each variable. You can make any combinations, i.e., 2 x 3 factorial design, 3 x 3 factorial design, and so on.

Calculating Factorial ANOVA

Example 8.2

3 x2 factorial ANOVA

If there are two independent variables in the experiment, a two-way factorial ANOVA would be used. There would be two main effects and one interaction. Main effects are tests of each independent variable when other is disregarded (controlled). Look at Table 8.6 and assume that the first independent variable (fiber) has three different fibers, labeled A1, A2, and A3. For example, A1 is oat bran, A2, wheat bran, and A3, rice bran. The second independent variable (fat) represents two different dietary fats: B1 is fish oil, and B2 is beef tallow. We can test fiber effects by comparing the row means (MA1, MA2, MA3) because dietary fat is equally represented at each level of A. That is for each level of A, two groups (fish oil and beef tallow) are included. Thus, dietary fat is held constant to allow the test of dietary fiber by the F_A ratio.

The same holds true for fats. By looking at the column means (MB1 and MB2), you can see that the three dietary fiber of A are equally represented in the two dietary fats of B. Therefore, the main effect of dietary fats can be tested by the F_B ratio.

In a study of this type, the main interest usually lies in the interaction. We want to know whether the effect of the dietary fibers of A depends on or changes across the dietary fats of B, that is whether the effect of dietary fiber depends on (interacts with) the dietary fats. This effect is tested by the F_{AB} ratio, which evaluates the six cell means: MA1B1, MA1B2, MA2B1, MA2B2, MA3B1, and MA3B2. Unless some special circumstance exists, interest in the testing of main effects is usually limited by the presence of a significant interaction, which means that what happens in one independent variable depends on the level of the other. Thus, normally it makes little sense to evaluate main effects when the interaction is significant.

This particular factorial ANOVA is labeled as a 3 (dietary fibers) x 2 (dietary fats) ANOVA. As the bottom of Table 8.7 indicates, the true variance can be divided into three parts:

Table 8.6

		Fats		
		B1	B2	Row means
Fiber	A1	A1B1	A1B2	MA1
	A2	A2B1	A2B2	MA2
	A3	A3B1	A3B2	MA3
Column means		MB1	MB2	

Fiber main effect is the test of the row means by F_A
Fat main effect is the test of the column means by F_B
The interaction is the test of the six cell means by F_{AB}.

F_A = (true variance due to A)/(error variance)
F_B = (true variance due to B)/(error variance)
F_{AB} = (true variance due to AxB)/(error variance)

*true variance due to A (dietary fiber)
*true variance due to B (dietary fat)
*true variance due to the interaction of A and B

Each of these true variance components is tested against (divided by) error variance to form the three F ratios for this ANOVA. Each of these Fs will have its own set of degrees of freedom so that it can be checked for significance in the F table in Appendix A.6. Table 8.7 gives the ABC method for calculating a two-way factorial ANOVA.

The ABC Method for Calculating a Two-Way Factorial ANOVA

Table 8.7

$A = \Sigma\, X^2$

$B = (\Sigma\, X)^2/N$

$C(\text{row}) = [\ (\Sigma\, Xr_1)^2 + (\Sigma\, Xr_2)^2 + \ldots\ldots (\Sigma\, Xr_i)^2]/n_{r1}$

$D(\text{column}) = [(\Sigma\, Xc_1)^2 + (\Sigma\, Xc_2)^2 + \ldots\ldots (\Sigma\, Xcj)^2]/n_{c1}$

$E(\text{rxc}) = [(\Sigma\, Xcell_1)^2 + (\Sigma\, Xcell_2)^2 + \ldots\ldots (\Sigma\, Xcell_k)^2]/n_{cell1}$

Summary table for ANOVA

Source	SS	df	MS	F
Rows (IV$_1$)	C - B	r - 1	SS_R/df_R	MS_R / MS_E
Columns (IV$_2$)	D - B	c - 1	SS_C/df_C	MS_C / MS_E
R x C	(E-B) - (C-B) - (D-B)	(r-1)(c-1)	SS_{RC}/df_{RC}	MS_{RC} / MS_E
Error	(A-B) - (E - B)	(N-1)-[(r-1)+(c-1)+(r-1)(c-1)]	SS_E/df_E	
Total	(A -B)	N - 1		

where r = number of rows
c = number of columns

Example 8.3

The effects of three dietary fats, beef tallow, soybean oil and fish oil, and two dietary fibers, oat bran and wheat bran on the increment of serum HDL cholesterol of male BHE rats were compared (not actual research data). Thirty rats were randomly assigned to one of the six groups (n=5), and they were fed respective diet for five weeks. Blood was collected by tail bleeding at the beginning of the study, and after a five-week experimental period and analyzed for HDL cholesterol. The increment of plasma HDL cholesterol was calculated, and the data are presented in the following table. Analyze the data and interpret the results.

		Dietary Fibers (IV_1)					
		Oat bran		Wheat bran			
		X	X^2	X	X^2	Σ	Mean
	Fish oil	20	400	16	256		
		19	361	17	289		
Dietary fats		21	441	15	225	181	18.1
		19	361	16	256		
		20	400	18	324		
	Mean	(19.8)		(16.4)			
	Soybean	15	225	16	256		
		14	196	17	289		
		13	169	15	225	152	15.2
		14	196	18	324		
		15	225	15	225		
	Mean	(14.2)		(16.2)			
	Beef tallow	13	169	15	225		
		12	144	14	196		
		11	121	16	256	135	13.5
		13	169	15	225		
		12	144	14	196		
	Mean	(12.2)		(14.8)			
	Σ	231		237			

$A = \Sigma X^2 = 7488$

$B = (\Sigma \ X)^2/N = (231 + 237)^2/30 = 7301$

$C(row) = [(\Sigma \ Xr_1)^2 + (\Sigma \ Xr_2)^2 + \ldots \ldots ( )]/10 = (32761 + 23104 + 18225)/10$
$= 7409$

$D(\text{column}) = [(\Sigma Xc_1)^2 + (\Sigma Xc_2)^2 + \ldots\ldots (\Sigma Xcj)^2]/n_{c1} = (53361 + 56169)/15$
$= 7302$

$E(\text{rxc}) = [(\Sigma Xcell_1)^2 + (\Sigma Xcell_2)^2 + \ldots\ldots (\Sigma Xcell_k)^2]/n_{cell1}$
$= (9801 + 5041 + 3721 + 6724 + 6561 + 5476)/5 = 7465$

$\text{Row} = C - B = 7409 - 7301 = 108$

$\text{Column} = D - B = 7302 - 7301 = 1$

$R \times C = (E-B) - (C-B) - (D-B) = (7465 - 7301) - (7409 - 7301) - (7302 - 7301)$
$= 55$

$\text{Error} = (A-B) - (E-B) = (7488 - 7301) - (7465 - 7301) = 23$

$\text{Total} = (A-B) = 187$

Summary table for ANOVA

Source	SS	df	MS	F
Rows	108	2	54	56.37
Columns	1	1	1	1.04
RxC	55	2	27.5	28.7
Error	23	24	0.958	
Total	187	29		

The test for dietary fats is significant, $F(2,24) = 56.37$ at both levels of $p<0.05$ (F=3.40) and $p<0.01$ (5.61), whereas the test for dietary fiber is not significant. However, the interaction effects are significant. Although the dietary fat effects are significant, we do not know whether all three groups differ significantly. Thus, we next perform a follow-up test. As mentioned in the Simple ANOVA, these methods include Scheffe, Newman-Keuls, Duncan's, and several others.

The Scheffe method is the most conservative, which means it identifies fewer significant differences. Duncan's test is the most liberal, identifying more significant differences. Newman-Keuls falls between the two. Therefore, we will use again Newman-Keuls test.

Newman- Keuls test

Group	M	2	3
1 Beef	13.5**	5.50**	14.89**
2 Soy	15.2**	9.39**	
3 Fish	18.1**		

1. Calculate error mean $E = \sqrt{MS_E/n} = \sqrt{0.958/10} = 0.309$
2. Order the means from highest to lowest: 18.1, 15.2, 13.5
3. Calculate the difference between the means: 2.9, 4.6, 1.7
4. Divide the difference by E and enter in table above: 2.9/.309=9.39, 4.6/.309=14.89, 1.7/.309=5.50
5. Calculate the steps between means; this is the number of means in the ordered set-- group 1 to 2 is two steps, groups 1 to 3 is three steps. Group 2 to 3 is two steps. The number of steps is called k.
6. The df for error is 24.
7. Enter the Studentized range table (Table A.7 in Appendix) with k and 24 df depending on which groups are being compared--- groups 1 and 2 would be 2 and 24 df ; groups 1 and 3 would be 3 and 24 df, and groups 2 and 3 would be 2 and 24 df.
8. Compare value calculated for table above with value in Table A.7 at alpha 0.05 (2.92 for 2 steps, and 3.53 for 3 steps), and 0.01 (3.96 for 2 steps and 4.54 for 3 steps).
9. *mark means that the calculated values are larger than the tabled value at $p<0.05$, indicating the differences are significant at $p<0.05$, and **mark indicates that the differences are significant at $p<0.01$ level.

Conclusion

Fish oil increased HDL cholesterol significantly more than either soybean oil or beef tallow, and soybean oil increased more than beef tallow. Therefore, fish oil may be the best dietary-fat source for the prevention of heart diseases (not real data).

Finally, the test for the interaction is significant, $F(2,24) = 28.7$, $p<0.01$. Thus,

149

Figure 8.1

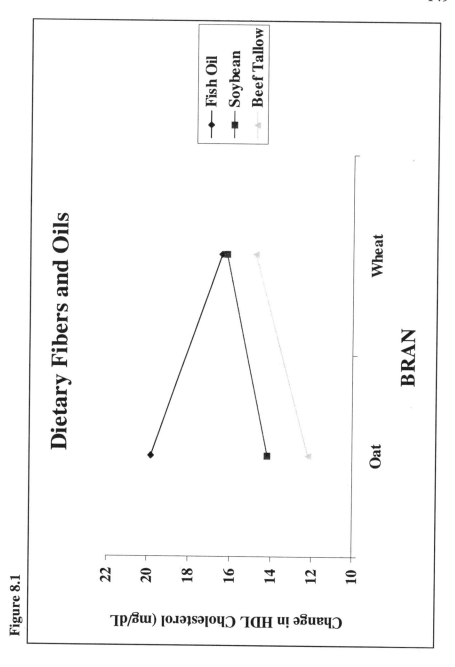

what happens with dietary fat depends on dietary fiber. **Figure 8.1** is a plot of this interaction. You can see that fish oil is the best, but it is the very best when consuming fish oil together with oat bran. In contrast, soybean oil and beef tallow are better when eating them with wheat bran than oat bran.

Exercise 8.2

Blood LDL cholesterol levels are as follows from the same experiment as the above HDL. Perform statistical analysis and interpret the results

		Dietary Fibers (IV$_1$)			
		Oat bran		Wheat bran	
		X	X^2	X	X^2
	Beef tallow	170		160	
		190		170	
Dietary fats		210		150	
		195		180	
		200		180	
	Soy bean	150		160	
		140		170	
		135		150	
		140		180	
		138		150	
	Fish oil	130		150	
		120		140	
		110		135	
		125		150	
		120		145	

$A = \Sigma X^2 =$

$B = (\Sigma X)^2/N =$

$C(row) = [(\Sigma Xr_1)^2 + (\Sigma Xr_2)^2 + \ldots\ldots (\Sigma Xr_i)^2]/n_{r1} =$

$$D(\text{column}) = [(\Sigma Xc_1)^2 + (\Sigma Xc_2)^2 + \text{.......} (\Sigma Xcj)^2]/n_{c1} =$$

$$E(rxc) = [(\Sigma Xcell_1)^2 + (\Sigma Xcell_2)^2 + \text{.......} (\Sigma Xcell_k)^2]/n_{cell1} =$$

Summary Table for ANOVA

Analysis of Covariance

Analysis of covariance (ANCOVA) is a combination of regression and ANOVA. The technique is used to adjust the dependent variable for some distractor variable (called the covariate).

In the two-way ANOVA, a test for the effect of one variable was made, separated from the effect of the second variable. The second variable was represented by several categories. If the second variable represents as actual measurement or score for each individual, we can again test for the effect of the first variable, separated from the effects of the second variable. The method of analysis to be presented is called the analysis of covariance. The second variable is often referred to as a "control" variable.

If we wish to compare the effects of different feeds on the weight of rats, a measurement of the weight of each rat before the experiment would be valuable as a control. If we wish to say that the AIN diet is the best, we should be able to state that the extra weight of the group fed by the AIN diet was not largely the result of the original weights. Even if the original weights are comparable, it is often impossible to have individuals in the experiment maintain equal amounts of intake of the particular diets. The related variable of total intake could be measured and taken into account in comparing the diets.

In an experiment designed to study the results of a dietary treatment to increase HDL cholesterol of three dietary fats we measure the HDL cholesterol, Y, of each individual at the end of the study and introduce the original HDL cholesterol, X, for each rat as a control variable. We may study the differences in the effectiveness for the three dietary groups with the use of Y variable, "controlled" or "adjusted" for the X variable.

The ANOVA procedure for difference in mean was based on the separation of a total sum of square into several portions. If the mean square for means was significantly large, we reject the hypothesis of equal means. The ANCOVA procedure also leads to a test for difference in means by separation of a sum of squares into several portions. In this case we test for a difference in means of "residuals." The residuals are the differences of the actual observations and a regression quantity based on the associated second variable.

Limitations of ANCOVA

Although ANCOVA may seem to be the answer to many problems, its use does have limitations. In particular, its use to adjust final performance for initial differences can result in misleading interpretations (Lord, 1969). In addition, if the correlations between covariate and dependent variable are not equal across the treatment groups, standard ANCOVA is inappropriate.

Summary

This chapter has presented techniques used in statistics in which differences among groups are the focus of attention. These techniques are categorized as follows.

A t test is used to determine how a group differs from population, how two groups differ, how one group changes from one occasion to next, and how several means differ (the multiple range test).

ANOVA shows differences among the levels of one independent variable (simple ANOVA), among the levels of two or more independent variables (factorial ANOVA), and among levels of independent variables when there is a distractor variable or covariates (ANCOVA).

References

Altman, D. Practical Statistics for Medical Research. London: Chapmen and Hall, 1991.

Boneau C.A. The effects of violations of assumptions undelying the t-test. Psychol Bull 1960; 57:49-64

Duncan D.B. Multiple range and multiple F tests. Biometrics 1955; 11:1-42

Fetters, N.L. The effects of a high and low intensity weight training program on bone mineral density in early postmenopausal women. Thesis, Master of Science, Norman, OK: University of Oklahoma, 1998.

Goldfried M.A. One-tailed tests and unexpected results. Psychol Rev1959; 66: 79-80

Green B.F. Tukey J.W. Complex analysis of variance: general problems. Psychometrika 1960; 25:127-152

Keuls M. The use of the "studentized range" in connection with an analysis of variance. Euphytica 1952; 1:112-122

Kirk, R.E. Experimental Design: Procedures for the Behavioral Sciences. (2nd ed.) Belmont, CA: Brooks and Cole, 1982.

Koh E.T. Effects of vitamin C on blood parameters of hypertensive subjects. Oklahoma State Med Assoc J 1984; 77:177-12

Lord E.M. Statistical adjustments when comparing preexisting groups. Psychological Bull 1969; 72:336-337

Newman D. The distribution of range in samples from a normal population, expressed in terms of an independent estimate of Biometrika 1989; 31:20-30

Sanders, D.H. Statistics; a First Course. (5th ed.) New York: McGraw-Hill, Inc., 1995.

Scheffe, H. The Analysis of Variance. New York: John Wiley, 1959.

"Student" The probable error of a mean. Biometrika 1908; 6: 1-25

Swinscow, T.D.V. Statistics at Square One. (9th ed.) British Medical Journal, 1996.

Thomas J.R. A note concerning analysis of error scores from motor-memory research. J Motor Behav 1977; 9:251-253

Thomas, J.R., Nelson, J.K. Research Methods in Physical Activity. Champaign, IL: Human Kinetics, 1996.

Tolson H. An adjustment to statistical significance: ω^2. Res Quart Exercise Sport 1980; 51: 580-584

Toothaker, L.E. Multiple Comparisons for Researchers. Newbury Park, CA: Saga, 1991.

CHAPTER 9

==

NONPARAMETRIC STATISTICS

Introduction

As discussed in Chapters 6, 7, and 8, statistical tests, such as the z, t, and F tests, are called parametric tests. Parametric tests are statistical tests for population parameters such as means, variances, and proportions that involve assumptions about the populations from which the samples were selected. One assumption is that these populations are normally distributed. But what if the population in a particular hypothesis-testing situation is not normally distributed? Statisticians have developed a branch of statistics known as nonparametric statistics or distribution-free statistics to use when the population from which the samples are selected is not normally distributed. In addition, nonparametric statistics can be used to test hypotheses that do not involve specific population parameters, such as μ, σ, or ρ.

Nonparametric statistics are versatile in that they can deal with ranked scores and categories. This can be a definite advantages when the investigator is dealing with variables that do not lend themselves to precise, interval-type data (that are more likely to meet parametric assumptions), such as categories of responses on questionnaires and various affective behavior rating instruments. Data from qualitative research are often numerical counts of events that can be effectively analyzed with nonparametric statistics.

Sometimes the only scores available are frequencies of occurrence or ranks (which often are not normally distributed), and the researcher should use nonparametric tests. For instance, "How often do you add salt to your food?" The researcher asked you to check one of the following: "seldom/never," "sometimes," "often/always" (see Survey questionnaire in Chapter 12). In this case, the researcher counts the frequency for each category, and uses nonparametric statistics. Several nonparametric methods are available. In this chapter, we describe a few of the more common ones.

Chi-square

The chi-square analysis (χ^2) is a commonly used nonparametric test. It is utilized for the data that are frequency counts in categories. Frequency counts could be the number of male and female, the number of subjects in the special age groups (20-29, 30-39, 40-49 etc.), and so on. With a chi-square analysis, observations are made and counted in one category. In this way the observations are said to be independent of one another. The premise behind the analysis is a comparison of what is expected to occur (theoretical probability) versus what is observed (empirical probability).

Empirical probabilities are derived from observations of certain events. Based on the frequency of occurrence, projections of future events can be made. Here is an example: Let's say that we have obtained and counted the number of men and women who had participated in our "Sex Hormone and Alcohol" study. Assume we have 200 male and 100 female participants so far. Based on these empirical observations, what would be the probability that the next person who participated in the study would be a man? The empirical probability would be $P = 200/300 = 0.67$, as 200 of the 300 participants in the past observations were men. The empirical probability for a women participating in the study would be $P = 100/300 = 0.33$. Of course, the theoretical probability for either a men or a women participating would be $P = 1/2 = 0.50$. The chi-square analysis then is a test to determine if there is significant deviation from what is expected to occur in a theoretical fashion.

One-way chi-square

The one-way chi-square is used to determine if there is a significant difference between the frequency of observed (empirical) and expected (theoretical) observations in two or more categories. The question for a one-way chi-square is

$$\chi^2 = \Sigma \frac{(O - E)^2}{E}$$

where O is the observed frequency and E is the expected frequency of a given category.

Example 9.1

Chi-square analyses for the above "Sex Hormone and Alcohol" study are shown in Table 9.1.

Table 9.1
Chi-square analysis

Category	Observed (O)	Expected (E)	O - E	$(O-E)^2$	$(O-E)^2/E$
Male	200	150	50	2500	2500/150=16.67
Female	100	150	- 50	2500	2500/150=16.67
Total	300	300			$\chi^2 = 33.34$

The resulting chi-square is then interpreted for significance by consulting Table A.8 in Appendix A. Note that the degrees of freedom for one-way chi-square is r - 1, where r is the number of categories. The chi-square sampling distribution is one-tailed, because it contains only positive values, just like F distribution. For this example there is one degree of freedom (2 - 1), and at the 0.05 level the critical chi-square is 3.84. The calculated chi-square of 33.34 is greater than 3.84. Therefore, we conclude that there is a significant difference between what was observed and what was expected. We reject the null hypothesis, and can conclude that there are gender differences for the numbers of participants, being more male participants than female participants in the study.

Example 9.2

We assume that nutrition researcher conducted a study entitled "the effects of soy on bone density in different ethnic groups." He wanted to determine the racial composition of his research subjects. His research technician reported that there might be a disproportionate number of four ethnic groups, Whites, Blacks, Asians, and Hispanics. Therefore, he counted the subjects for a month with the following results for 400 persons in those four ethnic groups. A summary of the analysis appears in Table 9.2.

Table 9.2 Chi-square analysis

Category	Observed (O)	Expected (E)	O - E	$(O-E)^2$	$(O-E)^2/E$
Black	105	100	5	25	0.25
Asian	90	100	10	100	1.00
Hispanic	85	100	15	225	2.25
White	120	100	20	400	4.00
Total	400	400			$\chi^2 = 7.50$

The degrees of freedom would be 4 - 1 = 3, and the critical value at the 0.05 level would be 7.82. Because the calculated chi-square of 7.50 is less than the critical value of 7.82, the null hypothesis is accepted. It can be concluded that there is no significant differences between what was observed and what was expected. Therefore, the researcher concluded that subject numbers were similar for all four ethnic groups.

Exercise 9.1

A fast-food restaurant is offering three new menu items, chili dogs (CD), soy-burgers (SB), and taco salads (TS). A day is selected at random, and the number of each new item ordered is tallied. The data follow. The manager hypothesized that there may be no preference in the menu selection. Perform chi-square analysis and interpret the results.

Item	CD	SB	TS
Frequency	57	68	43

Two-Way Chi-Square

When data are tabulated in table form in term of frequencies, a chi-square independence test can be used to test the independence of two variables.

Example 9.3

Suppose a new recipe for chicken soup is introduced to a number of students in Tulsa university. One can ask the question, "Do male students feel differently about this new recipe than female students, or do they feel basically the same way?" Note that the question is whether or not they prefer the new for chicken soup recipe but whether there is a difference of opinion between the two groups.

In order to answer this question, a researcher selects a sample of male and female students and tabulates the data in table form, as shown.

Table 9.3

Group	Prefer new recipe	Prefer old recipe	No preference	Row sum
Male	100	80	20	200
Female	50	120	30	200
Total	150	200	50	400

When data are arranged in table form for the chi-square independence test, the table is called a **contingency table.** The table is made up of R rows and C columns like ANOVA table.

Table 9.4

	Column 1	Column 2	Column 3
Row 1	$C_{1.1}$	$C_{1.2}$	$C_{1.3}$
Row 2	$C_{2.1}$	$C_{2.2}$	$C_{2.3}$

A contingency table is designated as an R x C table. In this case, R = 2 and C = 3; hence this table is a 2 x 3 contingency table. The degrees of freedom df = (R - 1)(C - 1). In this case $(2-1)(3-1) = 2$

Since the main question is whether there is a difference in opinion, the null hypothesis is:
H0: There is no gender difference on the opinion about the recipe of chicken soup.

The alternative hypothesis is:

HA: There is a gender difference on the opinion about the recipe of chicken soup.

To get the expected value for each cell, multiply the corresponding row sum by the column sum and divide by the grand total;

$$\text{Expected value} = \frac{\text{row sum x column sum}}{\text{Grand total}}$$

For each cell the expected values are computed as follows:

$$E_{1.1} = \frac{200 \times 150}{400} = 75$$

$$E_{1.2} = \frac{200 \times 200}{400} = 100$$

$$E_{1.3} = \frac{200 \times 50}{400} = 25$$

$$E_{2.1} = \frac{200 \times 150}{400} = 75$$

$$E_{2.2} = \frac{200 \times 200}{400} = 100$$

$$E_{2.3} = \frac{200 \times 50}{400} = 25$$

Table 9.5: Two way chi-square analysis

Category	Observed (O)	Expected (E)	$(O - E)^2/E$
Male/Pref.New	100	75	$(100 - 75)^2/75 = 625/75 = 8.33$
Male/Pref.Old	80	100	$(80-100)^2/100 = 400/100 = 4.0$
Male/No.Pref	20	25	$(20-25)^2/25 = 25/25 = 1.0$
Female/Pref.New	50	75	$(50-75)^2/75 = 625/75 = 8.33$
Female/Pref.Old	120	100	$(120-100)^2/100 = 400/100 = 4.0$
Female/No.Pref	30	25	$(30-25)^2/25 = 25/25 = 1.0$

$$\chi^2 = 26.66$$

The test is always a one-tailed right test, and the degrees of freedom are 2. If alpha = 0.05, the critical value is 5.99. Hence the decision is to reject the null hypothesis, since 26.66 > 5.99. Therefore, males and females differ in their opinions about the recipe. More males like new recipe, and more female like old recipe.

Exercise 9.2

A researcher wishes to determine whether there is a relationship between the gender of an individual and the amount of alcohol consumed. A sample of 100 people was selected, and the following data were obtained.

	Alcohol Consumption			
Gender	Low	Moderate	High	Total
Male	10	26	18	
Female	13	30	3	
Total				

At alpha = 0.05, can the researcher conclude that alcohol consumption is related to the gender of the individual?

Restrictions and Assumptions for Chi-Square

Certain conditions must be met in order to have a valid chi-square analysis. The primary assumptions and restrictions for its use include the following:

1. Data need to be frequency counts.
2. Observations need to be independent of one another.
3. Expected and observed frequencies need to equal each other.
4. Sample size needs to be adequate. A one-way chi-square with only two categories must have expected frequencies of at least five. When there are more than two categories for one-way or two-way tables larger than the 2 x 2, no more than 20% of the categories or cells may have expected frequencies less than five and none of the expected values may be less than 1. Also a two-way chi-square using 2 x 2 table cannot be used for sample size less than 20.

Spearman Rank Correlation Coefficient

In chapter 7, the techniques of regression and correlation were explained. In order to determine whether two variables are related, one uses the Pearson product moment correlation coefficient. Its values range from +1 to -1. One assumption for testing the hypothesis that $\rho = 0$ for the Pearson coefficient is that the population from which the sample is obtained is bivariate normally distributed. If this requirement cannot be met, one of the nonparametric equivalent calculation, the Spearman rank correlation coefficient (denoted by rs) or the Kendall correlation, can be used when the data are ranked.

The computations for the rank correlation coefficient are simpler than those for the Pearson coefficient and involve ranking each of the sets of data. The difference in ranks is found, and rs is computed by using these differences. If both sets of data have the same ranks, rs will be +1. If the sets of data are ranked in exactly the opposite way, rs will be -1. If there is no relationship between the rankings, rs will be near 0. The equation for the Spearman rs is:

$$\text{Spearman rs} = 1 - \frac{6 \Sigma d^2}{n (n^2 - 1)}$$

where d = difference in the rank; n = number of data pairs

Example 9.4

The nutrition majoring students baked eight different kinds of muffins using different fibers. Muffins were rated for texture by children and by adults. The scale ran from 1 to 20 points, with 1 being the lowest and 20 being the highest. The data are shown below. At alpha = 0.05 is there a relationship between the two ratings?

Muffin	A	B	C	D	E	F	G	H
Children	4	10	18	20	12	2	5	9
Adults	4	6	20	14	16	8	11	7

Rank each data set:

Muffin	Children	Rank	Adults	Rank
A	4	7	4	8
B	10	4	6	7
C	18	2	20	1
D	20	1	14	3
E	12	3	16	2
F	2	8	8	5
G	5	6	11	4
H	9	5	7	6

Compute d and d^2

X_1	X_2	$d = X_1 - X_2$	d^2
7	8	-1	1
4	7	-3	9
2	1	1	1
1	3	-2	4
3	2	1	1
8	5	3	9
6	4	2	4
5	6	-1	1
			$\Sigma d^2 = 30$

$$\text{Spearman } rs = 1 - \frac{6 \Sigma d^2}{n(n^2-1)} = 1 - \frac{(6)(30)}{8(8^2-1)} = 1 - \frac{180}{504} = 0.643$$

Find the critical value from Table A.9 in Appendix with n = 8 and alpha = 0.05. It is 0.738.

Do not reject the null hypothesis, since rs = 0.643, which is less than the critical value of 0.738. Therefore, it can be concluded that there is no significant correlation between the rankings of the children and adults.

Exercise 9.3

Six drinks were ranked by the male and female adults for flavor. The data are shown as the following (1 is the highest and 10 is the lowest ranking) Is there a relationship between the two rankings? Use alpha = 0.05.

Drinks	A	B	C	D	E	F
Male	1	5	3	7	9	1
Female	3	6	4	8	10	5

Some Other Nonparametric Tests

Mann-Whitney-Wilcoxon Test

The Mann-Whitney-Wilcoxon test is a nonparametric alternative to the t test for two independent samples. It is commonly used where experimenter draws two random samples from the same parent population, subjects each to a different experimental treatment, and compares the two on a single criterion. Of course, the experimenter may designate one of his samples as a control group, in which case that experimental treatment is no treatment. This test may also be used in situation where independent random samples are drawn from two different parent populations and compared on a single criterion to determine whether the two populations differ.

The Mann-Whitney-Wilcoxon test requires data on at least an ordinal scale, and this data is assumed to be continuously distributed. It does not require normality of distribution or homogeneity of variance for the groups under study. This is one of the most useful of the nonparametric tests; it is as powerful as t test under common research conditions, and the computations are quite simple.

Wilcoxon Matched-Pairs Signed Ranks Test

There are studies in which one has matched groups, or same group is tested twice on a dependent measure, and the investigator wants to know whether the difference between the matched groups (or the changes in scores for the same subjects) is significant. In such cases, the independent t test is not appropriate, and the researcher must use a dependent t test. The principle applies in nonparametric statistics. The researcher would not be correct in using a test such as the Mann-Whitney-Wilcoxon test with dependent (paired) groups.

The Wilcoxon matched pairs test is the nonparametric alternative to the correlated t and also requires that data be internal level. In the Wilcoxon test, the size of the difference between scores (such as between pre- and post test scores) is determined and ranked.

Kruskal-Wallis ANOVA

The analysis of variance uses the F test to compare the means of three or more populations. The assumptions for the ANOVA test are the populations are normally distributed and that the population variances are equal. When these assumptions cannot be met, the non-parametric Kruskal-Wallis test can be used to compare three or more groups. This test is sometimes called H test. It would be the statistic of choice when comparisons of more than two independent groups need to be made. Essentially all of the observation in the analysis are given a rank, then the sum of the ranks for each group is determined.

Friedman's ANOVA

Friedman's ANOVA is similar in theory to the Kruskal-Wallis test. It is used when more than two repeated measurements have been made on the same or related subjects. Therefore, Friedman's ANOVA is nonparametric version of the one-way ANOVA for repeated measures.

Summary

In this chapter we have examined a variety of nonparametric statistics that must be used when data are nominal or ordinal level or when the assumptions for parametric analysis cannot be met. There are several advantages to the use of nonparametric methods. The most important one is that no knowledge of the population distribution is required. Other advantages include ease of computation and understanding. The major disadvantage is that if the assumptions for parametric tests are met the nonparametric tests are less efficient than their parametric counterparts. In other words, slightly larger sample sizes are required to get as accurate a result as given by their parametric counterparts. The following are comparisons between nonparametric and parametric tests:

Table 9.8: Comparisons of statistical tests between parametric and nonparametric methods

Purpose	Parametric	Nonparametric
Compare the expected vs. observed frequency counts	None	Chi-square
Compare two independent samples	Independent t test	Mann-Whitney- Wilcoxon test
Compare two dependent samples	Dependent t test	Wilcoxon matched pairs test
Compare more than two independent groups	One-way ANOVA	Kruskal-Wallis ANOVA
Compare more than two related groups	One-way ANOVA repeated measures	Friedman's ANOVA
Relationship between two variables	Pearson r	Spearman r

References

Conover, W.J. Practical Nonparametric Statistics. New York: Wiley Inc., 1980.

Friedman M. The use of ranks to avoid the assumption of normality implicit in the analysis of variance. J Am Statistic Assoc1937; 32: 675-701

Harwell M.R. A general approach to hypothesis testing for nonparametric tests. Experimental Education, 1990; 58:143-156

Hollander, M., Wolfe, D.A. Nonparametric Statistical Methods. New York: Wiley, 1973.

Johnson, R. Elementary Statistics. (3rd ed.) London: Duxbury Press, 1980

Kendall, M. Rank Correlation Methods. London: Griffin, 1948.

Kennedy, J.J. Analyzing Qualitative Data: Introductory Loglinear Analysis for Behavioral Research. New York: Praeger, 1983.

Kerlinger, F.N. Foundations of Behavioral Research. New York: Holt, Rinehart and Winstone, Inc., 1958.

Kruskal W. Wallis, W. Use of ranks in one-criterion variance analysis. J Am Statistic A 1952; 47:583-621

Marascuilo, L.A., McSweeney, M. Nonparametric and Distribution-free Methods for the Social Sciences. Monterey, CA: Brooks and Cole, 1995.

Roscoe, J.T. Fundamental Research Statistics for the Behavioral Sciences. New York: Holt, Rinehart and Winston, Inc., 1969.

Siegel, S. Nonparametric Statistics for the Behavioral Sciences. New York: McGraw-Hill, 1956.

CHAPTER 10

===

MEASURING RESEARCH VARIABLES

Introduction

"In its broadest sense, measurement is the assignment of numerals to objects or events according to rules (Stevens)." This definition of measurement accurately expresses the basic nature of measurement. A numeral is a symbol of the form: 1,2,3,..., or I, II, III .... etc. There are four general levels of measurement: nominal, ordinal, interval, and ratio that we have already discussed in Chapter 6.

Comparisons of Scales: Practical Considerations and Statistics

In experimental-type research, measurement, defined as the translation of observations into numbers, is a vital action process in data collection. This action process links the researcher's theoretical concepts to concrete variables that can be empirically or objectively examined. The conceptual or theoretical consideration first involves identifying a concept and then defining it. For example, if you study effects of stress on blood pressure, first, you have to conceptualize or define "stress" and "blood pressure" in words. Then, the operational or empirical consideration of measurement involves "how to classify and/or to quantify what you observe." The strength of the relationship between indicator (measurement) and concept is critical. A measure is an empirical representation of an underlying concept. The more consistent or reliable and the more accurate or valid the relationship, the more desirable the measuring procedure or instrument. Thus, reliability and validity are two fundamental properties of indicators (DePoy & Gitlin).

The first step in the development of an indicator involves specifying how a variable will be operationalized. This process involves determining the numeric level at which a phenomenon can be measured.

The basic characteristics of the four types of measurement and their accompanying scales have been discussed (Chapter 6). What kinds of scales are used in scientific research? Mostly nominal and ordinal are used in applied research, while interval and ratio measurements are mainly used in basic research.

First, consider a great deal of measurement that is nominal. Measurements for the following cases, male or female, yes or no, married or single, children or no children etc., are nominal. With nominal measurement the counting of numbers of cases in each category and subcategory is, of course, permissible. For example, "male" may be assigned a value of 1, whereas "female" may be assigned a value of 2. The assignment of numbers is purely arbitrary, and no mathematical functions or ranking are implied or can be performed. Frequency statistics like χ^2, and percentages, certain measure of association can be used.

In the ordinal scale, measurement is obtained through some ordered ranking of groups in categories. However, the distance between each ranking is not known. A student who ranks second in the class is definitely higher than a student who ranks third, and lower than a student who ranks first. But we do not know the extent to which they differ. The statistics that can be used with ordinal scales include rank-order measures such as Spearman's correlation coefficient, Kruskal-Wallis ANOVA, median and percentiles. The lack of equal intervals is serious since distances within a scale cannot be added in a meaningful way without interval equality.

It is a goal of scientific measurement to construct and use appropriate scales. Once that goal is obtained, there is congruence between the properties being measured and the number system. In the state of measurement at present, however, we cannot be sure that our measurement instruments have equal intervals. The best procedure would seem to be treat ordinal measurements as though they were interval measurements. With sufficient knowledge of the subject matter and with sufficient care in the construction of instruments it is usually the case that measurement can be made at the interval level.

Reliability and Validity

After assigning numerals to objects or events according to rules, an investigator must face the two major problems of reliability and validity.

RELIABILITY

This refers to the degree to which an instrument accurately and consistently measures a given phenomenon from one time to another. An instrument with high reliability will yield comparable scores upon repeated administration. A high coefficient of reliability means a low error of measurement.

An instrument of measurement may be reliable even though it is not valid, but a valid instrument is always reliable.

There are several ways of computing the reliability of a measuring instrument. For some instruments the reliability has been established by the original designer. However, most instruments have no established reliability. Some of the common methods of assessing reliability include internal consistency, multiple form, test/retest, interobserver, and intraobserver.

Internal Consistency

This method measures the reliability of a self-reporting instrument (questionnaire, test) that is administered at the same time. In the split-half technique, instrument items are split in half, and a correlational procedure is performed between the two halves. In other words, the scores on one half of the test are compared with scores on the second half. The higher the correlation or relationship between the two scores, the greater reliability. A 0.8 correlation is interpreted as adequate reliability.

Multiple Form

In this approach, subjects respond during a single session to two instruments (or two forms of the same instrument) that measure the same fundamental quantity.

Suppose you wanted to know whether your client is feeling depressed. You ask, "How do you feel today?" The client answers, "Fine" and then begins to weep. However, you ask, "Do you feel depressed today?" you could have been more likely to obtain the desired information. For another example, suppose you want to know whether your subject drank alcohol last month. You ask, "Did you drink alcohol last month?" He may answer "No." However, he actually drank alcohol with his friends. Obviously he did not tell you "truth" because of his bad memory. On the other hand, you ask, "Did you drink alcohol last month?" "If yes, please indicate the type of setting in which you had alcohol:"

_____ Out with friend (club, bar, party)
_____ Dining out
_____ Dining at home
_____ Alone at a bar or club
_____ None

Then, he will remember and check at "Out with friends." You will get more reliable answer. The longer the questionnaire or more information collected to represent the underlying concept, the more reliable the instrument will become.

Test/Retest

This method measures the reliability of a self-reporting instrument that is administered at different time. The investigator wants to be assured that subjects will respond in the same way on repeated testing occasions. If the instrument yielded different scores each time a subject was tested, the scale would not be able to detect the "truth," or objective value, of the phenomenon under investigation. The longer the time span between the measuring times, the greater the likelihood of a low reliability.

Interobserver

This approach involves the observation of a phenomenon or performance by more than one person at the same time. If there are two or more observers involved, their performance can be rated and correlated. The total number of agreements between observers is compared with the total number of possible agreements and multiplied by 100 to obtain the percent of agreement. Eighty percent or more is considered to be an indication of a reliable instrument.

Intraobserver

This approach involves the observation of a phenomenon or performance at different times. If behavior is observed at two different measuring times, a comparison of the two observations can be made to test consistency and, thus, reliability. This measure is sensitive to changes in time. Because of concern over the impact of time change on an observed outcome, the interobserver approach is preferable to the intraobserver.

Systematic and Random Variability

There are two general types of variability: **systematic and random**. Systematic variability leans in one direction: scores tend to be all positive or all negative or all high or all low. Error in this case is constant or biased. Random or error variance is self-compensating: scores tend now to lean this way, now that way. Errors of measurements are random errors. They are the sum or product of a number of causes: the ordinary random or chance elements present in all measures due to unknown causes, temporary or momentary fatigue, fortuitous conditions at a particular time that temporarily affect the object measured or the measuring instrument fluctuations of memory or mood, and other factors that are temporary and shifting.

The principle behind the improvement of reliability is the one previously called the MAXICON principle (Chapter 4) in a slightly different form: "Maximize the variability of the individual differences and minimize the error variability." How to accomplish this principle in different areas will be discussed in the following.

To ensure the reliability of laboratory analysis, a comprehensive quality control is necessary. Every analysis in the laboratory generates a result. The validity of that result cannot be taken for granted, however, without some body of supporting evidence. The most convincing evidence is the establishment of a rigorous, comprehensive quality control program that is faithfully followed. This is the only way to guard against the possible deterioration of an analytic system, to become altered to imprecision of results when they occur, and to have confidence in test results when everything is in control.

Precision and Accuracy

The first step in the establishment of a quality control program is to ascertain the limits of uncertainty for each laboratory test. Every measurement, every analysis carries with it a degree of uncertainty, a variability in the answer as the test is performed repeatedly. It is essential to determine the precision of each test, which reflects the reproducibility of the test (the agreement of results among themselves when the specimen is assayed many times). The less the variation, the greater is the precision. Precision must not be confused with accuracy, which is the deviation from the true result. An analytical method may be precise but inaccurate because of a bias in the test method.

The degree of precision of a measurement is determined from statistical considerations of the distribution of random error; it is best expresses in terms of the standard deviation. A normal frequency curve is obtained by plotting the values from multiple analyses of a sample against the frequency of occurrence. From statistical considerations, the standard deviation, s, is derived from the following formula:

$$s = \sqrt{\frac{\Sigma (M - x)^2}{N - 1}}$$

where s = standard deviation, M = mean, x = any single observed value, and N = total number of observed values. With a normal distribution, 68% of the values are encompassed between $\mu - s$ and $\mu + s$, 95% between $\mu - 2s$ and $\mu + 2s$, and 99% between $\mu - 3s$ and $\mu + 3s$.

Example 10.1

We assume that we repeated vitamin D analysis ten separate times using a certain kit, and obtained the following data. To determine precision of the assay we should calculate standard deviation and coefficient of variation. Actually, the precision data for a test in the laboratory should be acquired from at least 20 separate assays. However, we use 10 assays here for a demonstration.

Assay values	M-x	$(M-x)^2$	
28	2	4	N= 10
30	0	0	M= 30
31	1	1	$\Sigma (M-x)^2 = 24$
27	3	9	
32	2	4	$s = \sqrt{\dfrac{\Sigma (M-x)^2}{N-1}} = \sqrt{\dfrac{24}{9}} = 1.6$
29	1	1	
30	0	0	
30	0	0	
31	1	1	
32	2	4	CV= coefficient of variation

$$CV = \frac{s}{M} \times 100 = \frac{1.6}{30} \times 100 = 5.33\ \%$$

In this example, 68% of the future assays are expected to fall within 30 ± 1.6 units of the mean value of 30, and 95% of the values within 30 ± 3.2 units from the mean, even when reagents, standards, and instruments are all acceptable, and there is no technologist error.

The standard deviation is greater when a method is less precise: it is generally greater when the mean of the assay values is a larger number. For example, the mean values of hemoglobin and standard deviation is 14 ± 2 (g/dL), whereas those of cholesterol is 150 ± 30 (mg/dL).

The coefficient of variation (CV) expresses the standard deviation as a percentage of the mean value and is a more **reliable means** for comparing the precision at different concentration levels. The precision of a method varies inversely with the CV; the lower the CV, the greater is the precision. The inter-assay CV less than 10 can be acceptable for the procedure, but it is better to be less than 5.

For instance, if intra- and inter-assay coefficients for osteocalcin are 3.9% and 4.5% whereas those of N-telopeptide are 6.5% and 7.9%, respectively, then the osteocalcin assay is more precise than that of N-telopeptide.

VALIDITY

The subject of validity is complex, controversial, and peculiarly important in scientific research. Validity as a concept of measurement refer to the degree to which a research procedure or tool measures what it is supposed to measure. In establishing validity measures there are six basic approaches used to obtain an accurate measure of the phenomenon under study. They are face validity, content validity, construct validity, concurrent validity, predictive validity, and criterion validity (Okolo).

Face validity refers to the degree to which an item in a measuring instrument "appears" to measure what it is supposed to measure. When developing or adopting a questionnaire, one must be careful and ensure that the items are specific enough to assess the desired variable. For example, a teacher has constructed a test to measure understanding of scientific procedures and has included in the test only factual items about scientific procedures. The test is not valid, because while it may reliable measure the students' factual knowledge of scientific procedures, it does not measure their understanding of scientific procedures.

Content validity refers to the degree to which the content of an instrument proportionately matches the preestablished objective criteria of a concept. For example, if the syllabus of diet- therapeutic course requires that the study spend one third of the time on cancer, at the end of the course one third of the evaluation (test) should cover cancer. If only one fourth of the test is on cancer, it will be low on content validity.

The problem with content validity two fold. First, for most concepts of interest to health and human service professionals, there is no agreed upon acceptance of the full range of content for any particular concept. For example, poverty, stress, depression, or self-esteem have been defined and conceptualized in many different ways. The second problems with this form of validity is that there is no agreed upon objective method for determining the extent to which a measure has attained an acceptable level of content validity.

Construct validity has to do with how one measure relates to other measures. It refers to the extent to which measures of a phenomenon have a pattern of correlations with other measures as an expected response or out come. For example, people with suicidal tendencies are expected to experience varied degrees of depression (lethargy, agitation, appetite changes, self-effacing thought, melancholia, etc.). Thus, the relationship of the scale score to other expected relationships would be measured as an indicator of construct validity.

Concurrent validity refers to the correlation between scores of two different scales or measures established at about the same time. For example, a student who has high SAT scores and a high grade point average at about the same time will establish a high concurrent validity. Students who score high on nutritional knowledge and a practice test and about the same time are able to select food in the cafeteria from the four basic food groups will establish a high concurrent validity.

In concurrent validity there is known standardized instrument that measures the underlying concept of interest. Suppose that you were interested in developing your own measure of stress. To establish concurrent validity, you would administer your instrument along with an already accepted and validated instrument measuring the same concept to the same sample of individuals.

Predictive validity means the extent to which scores on a scale or test predict future behavior. This can be measured by a method of correlation coefficients between individual score on a scale or test and later behavior. If students have high scores on a nutritional knowledge, attitude, and practice test and are later found to be interested and able to select food in the cafeteria from the four basic food groups, the nutrition scale is said to have high predictive validity.

Criterion validity can be either a concurrent criterion validity or a predictive criterion validity. Criterion validity is the process of comparing the cost effectiveness of the application of a costly instrument against a less costly one. In the correlation between the two instruments the criterion validity of the less costly instrument is considered high. Concurrent criterion validity refers to the use of the two instruments at the same time, while predictive criterion validity refers to using the less costly instrument at one time to predict the measures of the more costly instrument at the second time.

Practical Considerations

In addition to the issues of reliability and validity, the characteristics of subjects should be considered for data collection techniques and administrations. If one is interested in determining the accuracy of self-report of performance in self-care for older clients with mental illness, an interview may be appropriate with a closed-ended questionnaire (see Chapter 12). The characteristics of a sample also determine the appropriate length of an interview or contact time with the investigator.

The development of instrumentation will be discussed in the following Chapter.

Measurement Errors in Dietary Assessment

In the mid-eighteen century, observations that fresh fruits and vegetables could cure scurvy led Lind (1753) to conduct one of the earliest controlled clinical trials. They found that intakes of lemons and oranges were most effective in curing scurvy, which was ultimately found to be the result of vitamin C deficiency. Similarly, Goldberger used epidemiologic methods to determine that pellagra was a disease of nutritional deficiency, primarily associated with a corn meal subsistence diet in the southern United States.

Although these early observations demonstrated that diets could cause some diseases, the traditional methods of nutritionists did not address directly the relation between diet and occurrence of major disease of our civilization. Recently, the field of nutritional epidemiology has developed from interest in the concept that aspects of diet may influence the occurrence of human disease. Some epidemiological studies have explored the relationship between diet and certain chronic diseases (cancer, coronary heart disease, etc.), and made dietary recommendation to individuals and populations.

However, the serious limitation to the research in nutritional epidemiology has been the lack of practical methods to measure diet. Because such epidemiologic studies usually involve at least several hundred and sometimes tens of thousands of subjects, dietary assessment methods must be not only reasonably accurate but also relatively inexpensive. Accurate methods of dietary assessment may be obtained through minimizing errors in measurement in dietary intakes. To minimize errors, we should examine types of errors and their sources in measuring food and nutrient intakes.

Random and/or systematic errors may occur during the measurement of food and/or nutrient intakes. For random error, the average value of many repeated measures approaches the true value, that is, the law of large numbers applies. For systematic errors, the mean of repeated measurements does not approach the true value. In nutrition studies, random or systematic errors, or both can occur at two different levels: within a person and between persons. Thus, at least four types of error can exist: random within-person, systematic within-person, random between-person, and systematic between-person (Willett, 1998).

Random within-person error is typified by the day-to-day fluctuation in dietary intake. This apparently random variation is due both to the changes in food intake from day-to-day and to errors in the measurement of intake on any one day. Although it may be argued that true variation over time is not really error, it may be considered so if the long-term average intake for an individual is conceptually the true intake for that subject.

In addition to random error, repeated measurements of diet within a subject may also be subject to systematic error. This can occur for many reasons; in open-ended methods, such as 24-hour recalls, persons may consciously or unconsciously tend either to deny or to exaggerate their food intake. Systematic within-person error is particularly likely to occur when standardized questionnaires are used: an important food item for a subject (but not necessarily for all subjects) may have been omitted from a questionnaire or misinterpreted by a subject. If such a questionnaire is repeated, the same error is likely to recur;

thus the mean of many replicate measurements for an individual will not necessarily approach that person's true mean.

Random between-person error can be either the result of using only one or a few replicate measurements per subject in the presence of random within-person error or the consequence of systematic within-person errors that are randomly distributed among subjects.

Systematic between person error results from systematic within-person error that affects subjects nonrandomly. The mean value for a group of persons is thus incorrect. If the systematic error applies equally to all subjects and is simply additive, the observed standard deviation for the group is correct. If individuals, however, are affected to various degrees or the error is multiplicative, the observed standard deviation will also not represent the true standard deviation. Systematic between-person error are likely to be frequent and can have many causes: The omission of a commonly eaten food from a standardized questionnaire or the use of an incorrect nutrient composition value for a common food will affect all individuals in the same direction, but not to the same degree because the use of these foods will differ among subjects

The direction and extent of these errors vary with the method used and the population and nutrients studied.

Both types of measurement error can be minimized by incorporating a **variety of quality control procedures** during each stage of the measurement process. Random errors affect the precision of the method. They can be reduced by increasing the number of observations, but cannot be entirely eliminated. In contrast, systematic measurement errors cannot be minimized by extending the number of observations. They are important as they can introduce a significant bias into the results that cannot be removed by subsequent statistical analysis.

Now, we will discuss three different aspects: first, the sources of random and systematic errors which occur during the collection and recording of food consumption data; second precision in dietary assessment; and third, validity in dietary assessment methods.

Assessment and control of measurement errors

Random and systematic measurement errors can be minimized by incorporating various quality control procedures into each stage of the dietary assessment protocol. These can include training and retraining sessions for the interviewers and coders, standardization of interviewing techniques and questionnaires, pretesting of questionnaires, and administration of a pilot study prior to the survey.

Random errors, unlike systematic errors, can be minimized by increasing the number of observations. Random errors may occur across all subjects and all days. In contrast, systematic errors may exist for only certain respondents (e.g. obese, elderly subjects) and/or specific interviewers. The effects of systematic errors associated with specific interviewers can be minimized if it is ensured that the assignment of respondent to interviewer is random. In this way, subjects will not be questioned by the same interviewer on all occasions.

ERROR SOURCES

Respondent biases: The subject frequently tries to tell the researcher what he thinks the researcher wants to hear. Consumption of "good" foods such as fruits and vegetables may be over-reported, whereas the consumption of "bad" foods such as high fatty junk foods and alcohol may be under-reported. If the study population is stratified by relative weight, and overweight persons in the population tend to under-report their food intake, then a systematic bias in the mean intake for the overweight subgroup will occur (Anderson). The respondents frequently over-report their income and education levels, whereas they under-report the history of family diseases.

Interviewer biases may occur if different interviewers probe for information to varying degrees, and record responses incorrectly. The biases include errors caused by incorrect questions, incorrect recording of responses, intentional omissions, etc.

Respondent memory lapses will be minimized in recall methods if the time period between the actual food intake and its recall is short. Generally, foods contributing to the main part of a meal are remembered better than condiments, salad dressings.

Incorrect estimation of portion size of foods may be the largest source of measurement error in many dietary survey methods. The respondent may be unable to accurately quantify the portion of food consumed, or the perceived average serving may differ from the standard average serving. This type of errors can be reduced by using food models.

Supplement usage may be omitted from the dietary record or recall, causing significant errors in calculating nutrient intakes. If the researcher does not specifically ask about the single nutrient supplementation, it may not be reported.

Variation of nutrient contents in the same food may cause errors in calculating nutrient intakes. For example, intakes of selenium can be varied widely depending on the selenium content of soil where the food was produced; folic acid intakes can also be varied depending on the processing, cooking, and storage of folic acid-rich food because folic acid is very sensitive to those conditions.

Tendency to overestimate low intakes and underestimate high intakes in recall methods may be called the flat slope syndrome. The extent of the systematic error arising from the flat slope syndrome may vary with the age and sex of the study population. Over reporting could also be due to a food frequency questionnaire with an overly long list of foods. Under reporting could be due to failure of the researcher to ask the right questions about food preparation or use of a cultural food.

Coding and consumption errors arise when portion size estimates are converted from household measures into grams, or when food items are incorrectly coded (skim milk as whole milk, calcium-fortified orange juice as regular orange juice), and when foods are not listed in the data base.

Subject/researcher interaction errors: Since cooking methods affect nutrient intake, researchers must learn to ask the right questions regarding food preparation to obtain accurate answers. The same cooking term can mean different things to different people. Some cultures cook all of the food in one pot at the beginning of the day. After each meal, more ingredients are added to the pot for subsequent meals.

Random variation within subjects over time: The error lies with the strong variability in individual intake on a daily basis. Not only do the foods eaten on a daily basis vary, but also do the portion size. Activity levels and seasonal foods will also impact this variable.

Precision in Dietary Assessment

Epidemiologic studies attempting to relate usual dietary intake to disease use a variety of dietary instruments to gather information. In studies of cancer or other diseases in which the relevant exposure is past dietary intake, not current intake, the dietary history and food frequency are the most commonly used tools.

A questionnaire is reliable if it produces the same results when repeated a second time (assuming that the actual truth, here the dietary intake, hasn't changed). The reliability that is likely to be achieved by any instrument is influenced by the variability that it permits. For example, a questionnaire which does not permit variable portion size, or which uses categorical frequency response (e.g., 2-3 times/wk, 4-5 times /wk, etc.,), is likely to appear more reliable than one which allows for more variability in the responses. Thus, comparison of reliability of different questionnaires should be conducted with caution and with recognition of the different degrees of variability they may permit. Furthermore, reliability of the same instrument may be influenced by the method of administration, and particularly by the degree of error that the method of administration may foster. Interviewer administration is likely to appear more "reliable" than self- administration.

A dietary assessment method is considered precise (reliability/reproducibility) if it gives very similar results when repeatedly in the same situation. The precision is a function of the measurement error and uncertainty resulting from true variation in daily nutrient intakes.

1. Precision of 24-hour recalls

Single 24-hour recalls have been used to assess individual intakes, presumably on the assumption that the intake over one 24-hour period adequately represents the habitual intake. This assumption is unlikely to be correct for individuals living in most industrialized countries. Any estimate of an individual's usual intake, based on a single 24-hour recall, has low precision because of relatively large intra-subject variation in food intake. The precision can be improved by obtaining several 24-hour recalls for the same individual.

2. Precision in Food Record

To minimize errors resulting from memory lapses and inadequate estimation of portion size, a weighed food record is often used. A 7-day weighed record is usually considered appropriate for the estimation of average intakes of individuals. However, respondent burden is high and problems with compliance may rise. Consequently, shorter periods, ranging from 2- to 5-days, are often used.

3. Precision of Dietary Histories

The precision of the dietary history for assessing usual mean intakes of individuals and/or groups depends on the time frame of the method, the time lag of the method, the technique of measuring amounts of foods consumed, and the population group. In general, this method yields good precision when used for a group, especially over a relatively short time frame. On an individual level, the high intraclass correlation coefficients also indicated good overall agreement between the two dietary histories. For weekend days precision was poorer because of greater dietary variability at weekends.

4. Precision of Food Frequency Questionnaires

Acheson and Doll repeated a food frequency questionnaire after three months had elapsed. Respondents were asked to identify the frequency with which they consumed 48 foods and 8 types of drinks using a five point scale. After three months, 90% of the responses differed by less than one point from the original measurement, suggesting that the precision of the food frequency questionnaire was good. Similar results were obtained by Willett et al. after a time lapse of one year. The selection of food items must be careful for enhancements of the precision in calculating nutrient intakes.

Validity in Dietary Assessment Methods

Validity describes the degree to which a dietary method measures what it purports to measure. A valid assessment tool is a prerequisite for sound dietary assessment. If the instrument is biased or incomprehensive the data will be false, and the research will be failed. The errors that affect the validity of a dietary method are systematic error; those associated with precision are random.

1. Validity of 24-hour Recalls

The 24-hour recall can be validated directly, in terms of the actual intake, more readily than other methods of dietary assessment because the time frame covered is short. In general, the 24-hour recall tends to underestimate mean intakes in the elderly and children. Specific regarding food preparation methods are obtained, as they may affect nutrient values. Some studies have each subject complete multiple recalls, each on a different day of the week. It is important to include weekend intakes as most subjects have a higher caloric intake, especially of alcohol.

This is a completely inappropriate reference method for validating a diet history questionnaire. Because of day to day or intra-individual variability, 24-hr recalls do not represent the usual nutrient intakes of individuals very well, even for macronutrients. If carefully probed and quantitated, 24-hr recalls can produce reasonable group means. However, group means are not appropriate to use for validating a diet history questionnaire.

2. Validity of Dietary Histories

The dietary history method, developed by Burk in the 1940s, describes usual dietary habits and includes a detailed interview of usual daily intake and its variations plus a food frequency checklist and structured 3-day food record to serve as cross-checks. This method requires a highly trained nutritionist and is subject to intrainterviewer variation. Variations of the diet history that take less time to administer and focus on usual past intake have been developed primarily for use in large-scale epidemiologic studies. These methods generally collect information on the type of food eaten and the amount, frequency, cooking method, and addition.

The relative validity of the dietary history method has been most extensively assessed using the 7-day weighed or estimated food record as the reference method. In general, the dietary history produce higher estimates of group mean intakes than the 7-day record, especially if the estimated usual consumption for the dietary history has been assessed over a relatively long time period.

3. Validity of Food Records

Intermittent duplicate diet collections have been used to validate weighed and estimated food records. For example, in a U.S. study, 29 subjects consuming self-selected diets kept detailed weighed food records for one year and periodically made duplicate diet collections. The daily energy and nutrient intakes calculated from the one-year food records were significantly higher than those calculated from the records made during collection of the duplicate diets.

Several studies (Gibson & Scythes, Kim et al. Holbrook et al.) suggest that food intake changes during a duplicate diet collection period --- sometimes resulting in a decrease in energy intake of as much as 20%. Hence duplicate diets are not an ideal method for validating record methods.

For validity, careful review of needs and goals and instructions for accurate recordings are necessary. Then a follow-up review with the patient for portion sizes and methods of preparation, as well as brand names of products, prepares the record for nutritional analysis.

A study was conducted on approximately 250 middle-aged and older men (Sobell et al. 1989). Reference data consisted of 2-4 seven day records collected about two years apart, 12 to 15 years in the past. They were interviewed about their past diet using the National Cancer Institute questionnaire (Table 10-1). In the interview group, correlations for 15 nutrients were approximately 0.5 to 0.6 (not calorie-adjusted, with a few correlations approaching 0.7). In the group which received the questionnaire by mail, however, and in which there was no personal instruction nor querying of unreasonable responses, correlation with reference data were substantially lower; in the case of a few nutrients, they did not even exceed that which would be expected by chance alone. In both the interviewed and mailed groups, the group means estimated by the questionnaire were very close to those estimated by diet records. In the interview group, the correlations are essentially as good as can be obtained for current diet using diet-record reference data. Poor results in the mailed group indicate that careful attention to data quality is crucial to obtaining valid data.

4. Validity of Food Frequency Questionnaires

Food frequency questionnaires have become the primary method for measuring dietary intake in epidemiologic studies.

Food frequency, which is typically self-administered, asks the respondent to report usual frequency of consumption from a list of foods for a specified time period. Semi-quantitative food frequencies also ask questions on usual portion size. Some general questions may also be asked about cooking methods, type of fats used, and additions to foods.

Two well-known semiquatitative food frequencies are the Health Habits and History Questionnaire (Block) and the Nurses' Health Study Dietary Questionnaire (Willet). Recently, the performances of these two diet instruments were compared with a longer, interviewer-administered diet history (Caan et al.). Participants in a case-control study on diet and colon cancer were interviewed by trained nutritionists using validated diet history. Two separate subsamples of participants were asked to complete either the Block or the Willet questionnaire exactly 5 days after they completed the original diet history. Data were analyzed separately by subsample comparing either the Block or the Willet questionnaire with the original diet history by using means, correlations, quintile agreement, and odds ratios for the relation between several nutrients and colon cancer. The Block and the Willet questionnaires generally provided lower absolute intake estimates than did the original diet history; however, the Block questionnaire underestimated more than did that by Willet. Both correlations and quintile agreement were slightly better for the Willet questionnaire than for that by Block when compared with the original diet history. In general, point estimates obtained from either the Block or the Willet questionnaire fell within the confidence intervals of the estimates of the odds ratios obtained from the original diet history, and no real difference in significance levels appeared. They concluded that they were very similar in their ability to predict disease outcome although the Block and Willet questionnaires differed slightly from each other.

In constructing food frequency questionnaires, however, careful attention must be given to the choice of foods, the clarity of the questions, and the format of the frequency response section.

Various approaches that have been used to assess the performance of food frequency questionnaires, include the following:

- Comparison of means
- Proportion of total intake accounted for by foods included on the questionnaire
- Reproducibility
- Validity (comparison with an independent standard)
- Comparison with biochemical markers
- Correlation with a physiologic response
- The ability to predict disease

In general, diet records provide the best available comparison method; biochemical markers are potentially useful but do not exist for many dietary factors. Although multiple weeks of diet recording per subject provide the ideal standard, this process is costly. The use of a small number of replicate measures per subject with statistical correction for within-person variation provides an alternative approach that should make a validation study feasible in most epidemiologicsettings

Studies of absolute validity of food frequency methods are limited, despite their increasing use in epidemiological studies. So far based on prediction equations derived from food frequency questionnaires have had limited success in terms of their predictive ability and generalizability to other population group.

Use of Biochemical Markers to Validate Dietary Data

A biochemical markers defined as "any biochemical index" in an easily accessible sample that gives a predictive response to a given dietary component (Bingham).

However, like dietary intake, physiologic measures are also subject to measure-to-measure variability. Thus, it is rarely the case that a single serum or urine measurement actually represents long-term usual intake.

1. Doubly-labeled water technique

The doubly-labeled water technique is probably the closest to a "golden standard," but it only measures energy expenditure. In this technique, subjects drink a carefully measured dose of doubly-labeled water. Over the next 15 days, they must provide timed urine samples. Carbon-dioxide production and oxygen elimination are measured and a mathematical equation is used to determine energy expenditure. Changes in body weight and body water pool can be measured against the energy expenditure. This methodology is limited in mass epidemiological studies due to the high cost of the test and requirement of a mass-spectrometer that is used for measuring isotope.

2. Basal Metabolic Rate (BMR)

A less expensive method of group estimates is to obtain the subject's weight and estimate the basal metabolic rate (BMR). The BMR is then compared to the caloric intake from the multi-day dietary record. Caloric intake less than 1.2 times the calculated BMR should be excluded from the data, as this indicates the subject underestimated intake. If the BMR has actually been measured, then the exclusion is raised to 1.35. If only a 24-hour recall is obtained, then the exclusions are much lower at 0.92 at 95 percent confidence. Subjects could also be weighed before, during, and after the study. Any weight changes would point to instrument error (the record keeping causes the subject to alter normal eating habits) (Briefel, & Brown et al). These tests are not precise enough to validate individual intakes due to normal body weight fluctuations, individual variances of BMR as calculated from body weight, and variations in intake and energy expenditures.

3. 24-hour urinary nitrogen excretion

Isaksson was one of the first investigators to use nitrogen excretion levels in 24-hour urine samples to validate a 24-hour protein intake estimated by record or recall. This procedure is adopted because of the positive correlation observed between daily nitrogen intake and daily nitrogen excretion when dietary intake is kept constant. However, several factors can affect its reliability in estimating protein intake. People eat quite sporadically, which may cause protein intake vary. The subject must be also be in nitrogen balance, which would preclude those with muscle loss or repair due to starvation, injury, or intentional weight loss. Accurate 24-hour urinary nitrogen requires a complete urine sample. Incomplete samples result in underestimation of protein intake.

4. Complete 24-hour urine collections

Complete 24-hour urine samples are required. Creatinine excretion in the 24-hour urine samples is often used to measure the completeness of urine collections (Webster and Garrow). This validation of the completeness of the urine collection is based on the assumption that creatinine excretion is constant from day to day in an individual. To verify complete urine samples, the para-aminobenzoic acid (PABA) marker is used because PABA is a useful marker for monitoring the completeness of urine collection (Bingham and Cummings, 1983).

5. Urinary mineral excretion

Besides nitrogen, urinary excretion of certain electrolytes (e.g., Na^+, K^+, Ca^{++}) for which the urine is the major excretory route has also been used as a biochemical marker of dietary intake.

Anthropometric Assessment

The term "nutritional anthropometry" first appeared in "Body Measurements and Human Nutrition" (Brozek) and has been defined by Jeliffe (1966) as "measurements of the variations of the physical dimensions and the gross composition of the human body at different age levels and degrees of nutrition."

Anthropometric measurements are of two types: growth and body composition measurements. The body composition measurements can be further subdivided into measurements of body fat and fat-free mass, the two major compartments of total body mass. Anthropometric indices can be derived directly from a single raw measurement, weight for age, height for age etc., or from a combination of raw measurements such as weight and height, skinfolds thicknesses at various sites, and/or limb circumferences.

Anthropometric indices are of increasing importance in nutritional assessment as the measurement procedures have several advantages. The measuring procedures are simple, precise and accurate, provided that standardized techniques are used. The procedures can assist in the identification of mild to moderate malnutrition, as well as severe states of malnutrition. Relatively unskilled personnel can perform measurement procedures.

Measurement errors, both random and systematic, may occur in nutritional anthropometry. They arise from: examiner error resulting from inadequate training, instrument error, and measurement difficulties. Some of these measurement errors can be minimized by training personnel to use standardized, validated techniques, and instruments that are precise and correctly calibrated (Lohman et al.).

Clinical Assessment

Some physical abnormalities are diagnostic of nutrient deficiency, but most are mild and nonspecific. In fact, many signs can result from a lack of several nutrients as well as from non-nutritional causes. In spite of these shortcomings, their observation can be a useful index of nutritional status. Any positive findings should be further investigated or confirmed with anthropometric measurements, dietary evaluation or biochemical tests.

Because most clinical signs are nonspecific, they are best interpreted as part of a group of symptoms that are common to a particular nutrient deficiency. The greater the number of signs in a group are present, the greater the likelihood of a true nutrient deficiency.

Clinical examination procedures are simple, but the detection of clinical signs is very delicate. Therefore, clinical assessment should be performed by the experts in the area, or well-trained nutritionists.

Summary

Investigators face the two major problems of reliability and validity in measuring the research data. Reliability is the accuracy or precision of a measuring instrument. Validity is the ability of an instrument to measure what it purports to measure.

To ensure the reliability of laboratory analysis, a comprehensive quality control is necessary. Every analysis in the laboratory generates a result. The validity of the result cannot be taken for granted without some body of supporting evidence. The most convincing evidence is the establishment of a rigorous, comprehensive quality control program. The first step in the establishment of a quality control program is to ascertain the limits of uncertainty for each test. The degree of precision of a measurement is determined from the statistical considerations of the distribution of random error; it is best expresses in terms of standard deviation.

Dietary intake data in nutritional assessment systems have been severely criticized because of the random and systematic errors which occur in measuring nutrient intakes and because of the frequent misinterpretation of the data. This is unfortunate because the careful evaluation of nutrient intake data from target populations and /or individuals can produce important information, provided that limitations of the methods are clearly understood.

Random and/or systematic errors may occur during the measurement of food and/or nutrient intakes. The direction and extent of these errors vary with the method used and the population and nutrient studied. Both types of measurement error can be minimized by incorporating a variety of quality control procedures during each stage of the measurement process. These can include training and retraining sessions for the interviewers and coders, standardization of interviewing techniques and questionnaires, pretesting questionnaires administration of a pilot study prior to the survey. Random errors affect the precision of the method has been discussed. They can be reduced by increasing the number of observations, but cannot be entirely eliminated. In contrast, systematic measurement errors cannot be minimized by extending the number of observations.

References

Acheson E.D., Doll R. Dietary factors in carcinoma of the stomach: A study of 100 cases and 200 controls. Gut 1964; 5: 126-131

Anderson, S.A. Guidelines for Use of Dietary Intake Data. Life Sciences Research Office. Bethesda, MD: Federation of American Societies for Experimental Biology, 1986.

Beaton G.H., Milner J. Corey P., McGuire V., Cousins M., Stewart E., de Ramos M., Hewitt D., Grambsch P.V., Kassim N., Little J.A. Sources of variance in 24-hour dietary recall data: implications for nutrition study design and interpretation. Am J Clin Nutr 1979; 32: 2546-59

Bingham, S. "Biochemical Markers of Consumption." In the Dietary Assessment of Populations. Medical Research Council Scientific Report No 4:26-30, 1984.

Bingham S. Cummings, J.H. The use of creatinine output as a check on the completeness of 24-hour urine collection. Human Nutrition: Clinical Nutrition 1983; 39C:343-353

Block G., Hartman A.M., Dresser C.M. Carroll M.D., Gannon J., Gardner L. A data-based approach to diet questionnaire design and testing. Am J Epidemiol 1986; 124: 453-469

Bolland J.E., Yuhas J.A. Bolland T.W. Estimation of food portion sizes: effectiveness of training. J Am Dietet Assoc 1988; 88: 817-21

Briefel R.R. Assessment of the US diet in National Nutrition Surveys: National Collaborative Efforts and NHANES. Am J Clin Nutr 1994; 59(1S):164S-67S

Brown J.E., Buzzard I.M., Jacobs D.R., Hannan P.J., Kushi L.H., Barrosso G.M., Schmid L.A. A food frequency questionnaire can detect pregnancy-related changes in diet. J Am Dietet Assoc 1996; 96:262-6

Brozek J.F., Henschel A. (eds) Techniques for Measuring Body Composition. Washington, D.C.: National Academy of Sciences, National Research Council 1963.

Burk B.S. The dietary history as a tool in research. J Am Dietet Assoc 1947; 23: 1041-1046

192

Caan B.J., Slattery M.L., Potter J., Quesenberry C.P., Coates A.O., Schaffer D.M. Comparison of the Willet self-administered semiquantitative food frequency questionnaires with an interviewer-administered dietary history. Am J Epidemiol 1988; 148: 1137-1147

Comstock E.M., Symington L.E. Distributions of serving sizes and plate waste in school lunches. J Am Dietet Assoc 1982; 81: 413-422

DePoy, E., Gitlin, L.N. Introduction to Research. St. Louis, MO; Mosby 1991

DeVellis, R.F. Scale Development: Theory and Applications. Newbury Park, CA: Sage, 1991.

Gibson, R.S. Principles of Nutritional Assessment. New York & Oxford: Oxford University Press, 1990.

Gibson R.S., Scythes C.A. Trace element intakes of women. Brit J Nutr 1982; 48:241-48

Goldberger, J.E. Goldberger on Pellagra. Baton Rouge, LA: Louisiana State University Press, 1964

Hankin J.H., Stallones R.A., Messinger H.B. A short dietary method for epidemiologic studies. III Development of questionnaire. Am. J. Epidemiol 1968; 97:285-290

Heady J.A. Diets for bank clerks: development of a method of classifying the diets of individuals for use in epidemiological studies. J Roy Statist Soc 1961; A124:336-371

Holbrook J.T., Patterson K.Y., Bodner, J.E., Douglas L.W., Veillon C., Kelsey J.L. Mertz W., Smith J.C. Sodium and potassium intake and balance in adults ccnsuming self selected diets. Am J Clin Nutr 1984; 40: 786-93

Human Nutrition Information Service, US Department of Agriculture, Food Consumption: Households in the United States, Spring 1977, Washington, D.C: Government Printing Office, Publication H-1, 1982.

ICNND (International Committee on Nutrition for National Defense) Manual for Nutrition Surveys. (2nd ed.) Superintendent of Documents. Washington, D.C.: U.S. Government Printing Office 1963

Isaksson B. Urinary nitrogen output as a validity test in dietary surveys. Am J Clin Nutr 1980; 33: 4-6

Jelliffe, D.B. The Assessment of the Nutritional Status of the Community. WHO Monograph 53. Geneva; World Health Organization, 1966.

Kerlinger, F. Foundation of Behavioral Research. Holt, New York; Rinehart & Winston, 1973.

Kim W.W., Mertz W., Judd J.T., Marshall M.W., Kelsey J.L., Prather E.S. Effect of making duplicate food collections on nutrient intakes calculated from diet records. Am J Clin Nutr 1984; 40: 1333-37

Lohman, T.G., Roche, A.F., Martorell, R. eds. Anthropometric Standardization Reference Manual. Champaign, IL: Human Kinetics Books 1988.

Longnecker M.P., Chen M.J., Caan B. Block vs. Willet: A Debate on the validity of food frequency questionnaires (letter). J Am Dietet Assoc 1994; 94: 1: 16-19

Miller, D. Handbook of Research Design and Social Measurement. Newbury Park, CA: Sage, 1991.

Okolo, E.N. Health Research Design and Methodology. Boca Raton, FL: CRC Press, 1990.

Osgood, C.E., Suci, G.J., Tannenbaum, P.H. The Measurement of Meaning. Urbana, IL: University of Illinois Press, 1957.

Sobell J., Block G., Koslowe P., Tobin J., Andres R. Validation of a retrospective questionnaire assessing diet 10-15 years ago. Am J Epidemiol 1989; 130: 173-187

Spanier G.B. Measuring dyadic adjustment: new scales for assessing the quality of marriage and similar dyads. J Marriage Fam 1976; 38: 15-28

Stevens, S. "Mathematics, Measurement, and Psychologics." Handbook of Experimental Psychology. Stevens, S. ed., New York: Wiley, Inc., 1951.

Taylor, S. Validation of a Food Frequency Questionnaire Against a Three-Day Food Record. Master's Thesis, Oklahoma City, OK: University of Oklahoma 1997.

194

Webster J. Garrow J.S. Creatinine excretion over 24 hour as a measure of body composition of completeness of urine collection. Human Nutrition: Clinical Nutrition 1985; 39C: 101-6

Wechler, D. The Measurement of Adult Intelligence. (4th ed.) New York; Psychological Corp 1958.

Willett W., Sampson L. Stampfer M.J., Rosner B., Bain C., Witschi J., Hennekens C.H., Speizer F.E. Reproducibility and validity of a semiquantitative food frequency questionnaire. Am J Epidemiol 1985; 122:51-65

Willett, W. "Implications of Total Energy Intake for Epidemiologic Analysis." In Nutritional Epidemiology. McMahon, B. ed. New York: Oxford Univ Press, 1990

Willett, W. Nutritional Epidemiology. (2nd ed.) New York & Oxford: Oxford University Press 1998.

Zung W.K. A rating instrument for anxiety disorders. Psychomatics 1971; 12: 371-379

PART III

==

TYPES OF RESEARCH
&
EXPERIMENTAL DESIGN

Part III explores the various types of research: experimental, quasi-experimental, descriptive, and qualitative research. We also discuss the various types of experimental designs.

Chapter 11 introduces experimental research. We discuss the various types of internal and external validity, and the strength and weakness of various experimental designs: pre-experimental, true experimental, and quasi-experimental.

Chapter 12 discusses differences in descriptive and qualitative research. Survey research is the most common type of descriptive research in dietetic, nutrition, and health areas. Survey typically use four general procedures for data collection – the questionnaires, the personal interview, Delphi method, and the normative survey. This chapter covers types of nutrition descriptive and epidemiological descriptive research. We discuss qualitative research technique, which are mainly used in the ethnographic cultural studies.

CHAPTER 11

===

EXPERIMENTAL AND QUASI-EXPERIMENTAL RESEARCH

Introduction

Research design is the plan, structure, and strategy of investigation conceived so as to obtain answers to research questions and to control variance. Research design has two basic purposes: (1) to provide answers to research questions and (2) to control variability. The main technical function of research design is to control variance: **Maximize systematic variability, control extraneous systematic variability, and minimize error variability.**

Experimental research attempts to establish cause-and-effect relationships. That is, an independent variable is manipulated to judge its effect upon a dependent variable. However, the process of establishing cause and effect is a difficult one. We have already discussed the fact that just because two variables are correlated does not mean one causes the other. However, cause and effect cannot exist unless two variables are correlated. Therefore, correlational research is frequently used before conducting experimental research. We may investigate to see whether two variables are related before trying to manipulate one to change the other.

Also remember that cause and effect are not established by statistics. Statistical techniques can only reject the null hypothesis (established that groups are significantly different) and identify the percent variance in the dependent variable accounted for by the independent variable or the effect size; neither of these procedures establishes cause and effect. Cause and effect can be established only by the application of logical thinking to well-designed experiments. This logical process establishes that no other reasonable explanation exists for the changes in the dependent variable except the manipulation done as the independent variable.

In this chapter we will discuss experimental designs by explaining how you can recognize and control sources of invalidity and threats to both internal and external validity. In Chapter 10, the validity or accuracy of an instrument or test was discussed. In this Chapter we will discuss other type of validity which pertains to the outcome of an experiment or study.

Internal and External Validity

Two concepts, external and internal validity, are vital to a conceptual grasp of all research design. Use of the term "validity" in this context is somewhat different from its use in test and measurement areas (Chapter 10). In the context of experimentation, Campbell and Stanley defined as follows: "Internal validity is the basic minimum without which any-experiment is uninterpretable." An experiment that is internally valid is characterized by having successfully controlled all systematic influences between groups being compared, except the one under study. For example, a researcher will compare the effect of animal protein and plant protein on children's' intelligence quotient (IQ) development. For this experiment, it would be desirable to have the only systematic difference between the groups be protein source. Any other systematic differences between the two groups may result in IQ differences that appear to be the result of protein source but really would not be. Consequently, to design an internally valid experiment, the researcher must eliminate all systematic differences between two groups except protein source.

External validity, as described by Campbell and Stanley, "asks the question of generalizability: To what populations, settings, treatment variables, and measurement variables can results obtained be generalized?" For example, if the results of a study showed that fish oil lowered cholesterol for men, can it work the same for women? The ability to extend the results of a study into other situations is called "generalizability." The greater the ability to generalize the results of a study, the higher the degree of external validity it has.

Internal Validity

All types of experimental designs have some strengths and weaknesses that pose threats to the validity of the research. First, we will discuss several threats to internal validity.

The independent and dependent variables have been discussed. For quick review: The independent variable is the one that the researcher manipulates. It is the variable that is referred to as the treatment, the one that is tested to determine if it caused an effect. The dependent variable is the one that is affected or is the one that the investigator is observing a change in. Internal validity, then, is the extent to which the research condition is controlled so that the independent variable causes an effect or change in the dependent variable. The stronger the control over an experimental condition the higher degree of internal validity. Unfortunately there is no statistical value associated with internal validity for the researcher to judge. So one must carefully examine how the research was conducted and draw one's own qualitative conclusion.

Many factors have been identified as key threats to the internal validity of a study. Campbell and Stanley reported on these factors in their classic book, which has been presented in virtually every research textbook since then. A discussion and illustration of some of these threats follow. As each of the factor is discussed, it will be helpful to recall exactly what is of concern. These threats are conditions or things that may cause a change in the dependent variable other than the independent variable.

Campbell and Stanley identified eight threats to the internal validity of experiments:

1. History

History relates to all of the things that occur outside of a study to the subject that may influence the dependent variable. Typically these factors are recognized as possibilities but can't be controlled for. For instance, perhaps a nutritionist is interested in examining the effect of a special diet on weight loss among entering college freshmen. Several students in the study are incoming members of the same sorority in where most of the current members are much concerned about body image and thinness, and anorexia nervosa is highly prevalent among the members. This may have created a condition of increased peer pressure to lose weight. At the end of the study it would be difficult to conclude if the diet or the peer pressure may have caused weight loss. Of course, there could be many other such outside influences. Controlling the effects of history (peer pressure here) can be very difficult.

2. Maturation

Distinctive from history problems, maturation refers to factors that influence subjects' performance because of time passing rather than specific incidents. Subjects changes in many ways as a result of time passing. Included in Campbell and Stanley's discussion of maturation influences are such examples as growing older, hunger, and fatigue. Any time growth, learning or maturation processes occur during the conduct of a study that influence the dependent variable, it raises the uncertainty of the effect that the independent may have had. Maturation effects are the most concern when children are the subjects because of their rapid or unpredictable growth or changing rates. Especially, this threats is of concern in longitudinal research such as is conducted in child development, in which the actual physiological or mental growth process is involved. For instance, an investigator studies the effect of fish oil on IQ in young children for two years. At the end of the study it would be difficult to conclude whether fish oil or the natural growing of IQ in children may affect their IQ development.

3. Testing

The process of testing provides a subject with some experience about the test that may influence subsequent test performance. For example, if a group of nutrition students were administered a 50 multiple-choice test to evaluate their nutrition knowledge today and again one week later, the students would do better the second time even though no education involved. The testing experience may also provide an increased incentive to improve in some manner that is **independent** of the experimental treatment. The threat of test practice is most obvious in investigations in which the researcher administers a pretest, a treatment, and then a posttest on the same subjects. Presumably, the researcher is assessing the difference between pretest and posttest performance with the intent of interpreting differences as being caused by the treatment. However, a portion of the change from pretest to posttest performance levels may be caused by test practice.

4. Instrumentation

"Instrumentation" has been defined as "changes in the calibration of a measuring instrument or changes in the observers" that may influence the score or measures obtained (Campbell and Stanley). Instrument inaccuracies can occur due to inappropriate use or other violations in sound measurement. Suppose the researcher uses a spring-loaded device to measure bone strength. Unless the spring is calibrated regularly, it decreases in tension with use. Thus, the same amount of applied force will produce increased reading of strength. Also, the question of instrument accuracy is confounded when more than one person is involved in measuring bone strength. For another instance, interviewer biases may occur if different interviewers probe for information (e.g. nutrient intakes) to varying degrees (Chapter 10).

5. Regression to the Mean

Regression to the mean may occur when groups are not randomly formed but are selected on the basis of atypical, extreme score on some measure. For instance, the phenomenon of statistical regression occurs when an individual scores high in an initial testing situation, then tends to score lower in a subsequent test. The reverse holds true for the initial lower score, who tends to score higher on the next. So for individual or group that had a low pretest score, a legitimate question becomes: Did the pretest or retest score improves because of the independent variable or because of regression to the mean? This is important point for the researcher to consider. Further, regression to the mean is usually more of a problem with small sample sizes.

6. Selection Bias

Selection biases occur when groups are formed on some basis other than random assignment. This is most likely to occur when volunteers are free to select control group or treatment group. The bias of their option may indicate a stronger sense of motivation or predisposition than that of someone who did not select the control or treatment. This is especially confounding when subjects are asked to volunteer to be in a certain group. It is difficult to make a conclusion whether the group differences were due to any original discrepancies or due to experimental treatments.

7. Experimental Mortality

Experimental mortality refers to the loss of subjects from the experimental groups. Subjects may drop out of an experimental group for many obvious reasons. Here are examples of our seven-month study entitled, "the effect of a high and low intensity weight training program on bone mineral density in early postmenopausal women." Thirty-six subjects originally participated in the study, however, due to injury, time commitment, job relocation, and compliance standard ten subjects discontinued and/or were eliminated from the study. There were four individuals who dropped out of the low weight training group due to acquired illness or time commitment. Three subjects were excluded from the high weight training group because of injuries (not related to the training) or time commitment. Two of the controls were excluded due to the inability to schedule post testing and third control subject was removed because of non-compliance.

Then, what type of problems does mortality cause? Subjects were assigned in an unbiased manner at the beginning of a study, it may be concluded that by and large they are similar to one another. After subject attrition occurs, the remaining subjects may be unique from the standpoint of motivation, health, interest, and the like. What then remains may be dissimilar groups. So this creates essentially the same problem that selection biases caused. We do not know if post study differences can be attributed to the unique attributes of the remaining subjects or because of the independent variable.

8. Selection-Maturation Interaction

A selection-maturation interaction occurs only in specific types of designs. In these designs, one group is identified because of some specific characteristic that is being studied that will improve naturally over the passage of time (e.g., physical injury), whereas the other group lacks this characteristic. For instance, a researcher is investigating the effects of vitamin C on relieving from the cold symptoms. There are three experimental groups: 1000 mg vitamin C supplemental group, 500 mg supplemental group, and the control group. If we assume that 1000 mg supplemental group get relieved from the cold symptoms faster than the other two groups. In this case it would be difficult to know whether the independent variable, vitamin C, caused the change or the condition in 1000 mg supplemental groups improved by itself faster than that of the other two groups. This threat is compounded when groups are formed on the bias of this characteristics, which is a form of selection bias, and results in nonequivalent groups.

The threats to validity just described are the classic ones presented by Campbell and Stanley. Four other points should be considered as follows:

9. Placebo Effect

It has been observed that mere participation in a study may elicit an effect that is separate from the independent variable. In medicine, control subjects who participate in a drug study are given a placebo, which is simply an inactive substance. For instance, we gave calcium gluconate as a placebo to control group, whereas vitamin C (1,000 mg/day) was given for the treatment group (Koh). Calcium gluconate and vitamin C were similarly coated so that the subjects were not able to differentiate between the two. To avoid tipping off the subjects, drugs and placebos are usually administered in what is called a blind fashion, which prevents subjects from knowing what treatment they are receiving.

10. Halo Effect

The halo effect usually is introduced when the researcher has some expectation about the performance of a subject and is assessing the experimental variable. For instance, a physician studies the effects of sugar on children's' hyperactivity. If the physician expected children from the sugar group to be more active for some reasons, he may look for more signs of activity in that group, thus biasing the observations and assessments. Hopefully, you can see why it is important for the researcher to be as objective as possible. To reduce this halo threats, double-blind experimental design should be used.

11. Hawthorne Effect

Hawthorne effect was first recognized by studies done of workers at the Hawthorne Plant of the Western Electric Company in Chicago years ago. Basically, the investigators were interested in studying the relationship between brightness of light and work output. It seemed that as the lights were made brighter, work output increased. After a certain point, the investigators reduced the lighting intensity to see if the relationship continued. To their surprise, as the illumination decreased, the workers' productivity continued to increase. The investigators concluded that the increase work output was probably because of attention the subject received. Therefore, because subjects know they are being studied, they try harder. The Hawthorn effect definitely a concern when conducting an experimental study in which groups are compared.

12. Avis Effect

Another threat to internal validity wherein subjects in the control group may try harder just because they are in the control group. To reduce this effect the subjects should not be told what treatment they receiving. Avis effect can be eliminated by using blind method.

External Validity

Like internal validity, external validity has no statistical references and is qualitatively determined by the researcher. Campbell and Stanley presented the following threats to external validity:

1. Reactive or Interactive Effects of Testing

Pretesting a subject may enhance performance on retest, or alter subject's perception about the experimental treatment. Their perception may be changed in such a way that they are motivated to perform better than without pretest. When this occurs it can be said that there is a reactive or interactive effect that the testing experience has with the experimental treatment. Therefore, the pretest threatens external validity in a similar but slightly different conceptual context than learning to respond. The pretest may result in an increase or decrease in subjects' sensitivity to the variable under study. If this occurs, the results cannot be generalized to those who are not pretested.

2. Interaction of Selection Bias and Experimental Treatment

It can be referred as population-sample difference effect. One obvious dimension of external validity concerns the degree to which those subjects in the study are representative of the population to which generalization is desired. If there are significant differences between subjects and the larger population, it is likely that the subjects will respond differently than the larger group would. In this case, the results from the subjects cannot apply to the population.   For instance, a nutrition researcher was interested in examining the effect of a program designed to improve dietary adherence, and he obtained the results through testing the program with thirty nutrition majoring students. Therefore, the effectiveness of the program and generalizability of this study are limited to generalize only to the students majoring in nutrition, but not to the whole college students.

3. Reactive Effects of Experimental Arrangements

A third threat to external validity is embodied in the actual research arrangements chosen for the investigation. The question is whether a subject will perform in the same manner under experimental conditions as he or she would be in the real world. To the degree that the research setting deviates from the real world, it may be expected that the subject's response is changed. In one sense this phenomenon is the Hawthorne effect operating as a threat to the external validity. The more artificial and constrained a research setting becomes, the less the ability to generalize from it. Therefore, the result of the laboratory experiment are less ability to generalize than that of field study.

4. Multiple Treatment Interference

The influence of multiple treatment interference as a threat to external validity is conceptually similar to that of pretesting. It becomes operative when more than a single treatment is administered to the same subjects. Multiple-treatment investigations present obvious analytical internal validity problems, since beyond the first treatment, a subject's performance is difficult to attribute to any given treatment (e.g., treatment 2 or 3) because of the cumulative effect. Additionally, however, the generalizability of results from treatments subsequent to the first is also questionable. Because the effects of prior treatments are often not dissipated by the time later treatments are administered, the subjects may well have a depressed or increased sensitivity to a second or third research task. To the degree that such subjects' responsibility is altered for, say, treatments 2 or 3, the generalizability of results from these treatments is reduced.   The performance of one task may inhibit or enhance the performance of an opposing one. For instance, perhaps performing a motor skill test for balance will allow someone to do better on a subsequent balance test.

Controlling Threats to Internal Validity

Threats to internal and external validity are controlled in different ways and by specific techniques. In this section we describe useful approaches to solving these problems in the design of experiments. Many threats to internal validity are controlled by equating the subjects in the experimental and control groups. This is most often done by randomly assigning subjects to groups.

Randomization

As mentioned in Chapter 6, randomization allows the assumption that the groups do not differ at the beginning of the experiment. The randomization process controls for history up to the point of the experiment; that is the researcher can assume that past events are equally distributed among groups. It does not control for history effects during the experiment if experimental and control subjects are treated at different times or places. Only the researcher can make sure that no events occur in one group but not in others.

Randomization also controls for maturation, as the passage of time would be equivalent in all groups. Statistical regression is controlled because it operates only when groups are not randomly formed. Both selection biases and selection-maturation interaction are controlled because these threats occur only when groups are not randomly formed.

Sometimes ways other than random assignment of subjects to groups are used to attempt to control threats to internal validity. The matched-pair technique equates pairs of subjects on some characteristic and then randomly assigns the pairs to groups.

In within-subjects designs, the subjects are used as their own controls. This means each subject receives both the experimental and control treatment. In this type of design, the order of treatments should be counterbalanced; that is, half the subjects should receive the experimental treatment first and then the control, and the other half should receive the control first and then the experimental treatment such as crossover design.

Placebos, Blind, and Double-Blind Setups

Other ways of controlling threats to internal validity include placebos and blind and double-blind setups. A placebo is used to evaluate whether the treatment effect is real or a psychological effect. Frequently, a control condition is used in which subjects receive the same attention from and interaction with the experimenter, but the treatment administered does not relate to performance on the dependent variable.

A study in which subjects do not know whether they are receiving the experimental or control treatment is called a blind setup. In a double-blind setup, neither the subjects nor the tester knows which treatment the subjects are receiving. The triple-blind test has also been reported (Day). The subject does not know what he or she is getting. Moreover, neither the experimenter nor the investigator knows what the subject is getting.

All these techniques are useful in controlling Hawthorne, Halo effect, and Avis effects.

Uncontrolled Threats to Internal Validity

Three threats to internal validity remain uncontrolled by the randomization process. They are reactive or interactive effects, instrumentations, and experimental mortality.

Reactive or Interactive Effects

Reactive or interactive effects of testing can be controlled only by eliminating the pretest. However, it can be evaluated by two of the designs: randomized groups pretest/posttest and Solomon four groups that we will discuss in the section of research design.

Instrumentation

Instrumentation cannot be controlled or evaluated by any design. Only the experimenter can control this threat to internal validity. In Chapter 10 we discussed some techniques for controlling the instrumentation threat (developing valid and reliable tests). Of particular significance is test reliability. This frequently involves the assessment of test reliability across situations, between and within testers or observers, and within subjects. The validity of the instrument (does it measure what it was intended to measure?) must also be established to control for instrumentation problems.

Experimental Mortality (Loss to Follow-up)

Experimental mortality is not controlled by any type of experimental design. Only the experimenter can control this by ensuring that subjects are not lost from groups. Many of these problems can be handled in advance of the research by carefully explaining the research to the subjects and the need for them to follow through with the project. Monetary rewards after completion of the study will be helpful for the subjects to retain in the study.

Controlling Threats to External Validity

External validity is generally controlled by selecting the subjects, treatments, experimental situation, and tests to represent some large population. Of course, random selection is the key to controlling most threats to external validity. Not only the subjects are randomly selected, but also the levels of treatment can be randomly selected from the possible levels. The dependent variable could be randomly selected from a pool of potential dependent variables.

If the experiment is conducted under controlled laboratory conditions, then the findings may apply only under the controlled laboratory conditions. Frequently, the experimenter hopes the findings will generalize to real-world dietary, nutritional, and health settings. Whether the outcomes will generalize in this way depends largely on how the subjects perceive the study, and this influences the way subjects responded to study characteristics. The question of interest here is: Does the study have enough characteristics of real-world settings so that subjects respond as if they were in the real world? This is not an easy question to answer. Applied research, e.g., dietary surveys, nutrition surveys, and epidemiological studies, are mainly conducted in field settings, whereas most of basic research, e.g., clinical, and animal studies, are mainly controlled laboratory settings.

Reactive or interactive effects of testing can be evaluated by the Solomon four group design. Interaction of selection biases and the experimental treatment is controlled by random selection of subjects. Reactive effects of experimental arrangements can be controlled only by the researcher. Multiple-treatment interference may be partially controlled by counterbalancing or randomly ordering the treatments across the subjects. But only the researcher can control whether the treatments will still interfere. That decision is based on knowledge about the treatment rather than the type of experimental design.

Experimental Research

Experimental research designs are classified by their sophistication and degree of control they provide. The categories of these designs are called true, quasi-, and pre-experimental. One thing all experimental designs have in common is the introduction and manipulation of an independent variable. There are many possible research design to choose from. The selection of design is based on factors like the type of research problem, the number and variety of experimental variables, how groups will be formed, the threats to validity the researcher wants to control for, and so on.

1. Pre-Experimental Designs

The pre-experimental designs are weaker than the true experimental designs in terms of control. Pre-experimental designs have no random sampling of subjects, are usually one group or two unequated groups, control few threats to validity, and have many definite weaknesses. Therefore, the pre-experimental designs are not discussed in detail here.

2. True Experimental Designs

The true experimental design is perhaps the design best known by beginning researchers and lay persons. True experimental designs refer designs similar to the classic two group design in which subjects are randomly selected and randomly assigned (R) to either an experimental or control group. Before the experimental treatment, all subjects are pretested or observed on a dependent measure (O). In the experimental group, the independent variable or experimental treatment is imposed (T), and it is withheld in the control group. Subjects are then post-tested or observed on the dependent variable (O) after experimental condition (Thomas & Nelson).

In this design, the investigator expects to observe no difference between the experimental and control groups on the dependent measure at pretest. That is because subjects are chosen randomly from a larger pool of potential subjects and then assigned to a group on a chance-determined basis, subjects in both groups are expected to perform similarly. On the other hand, the investigator hypothesizes that differences will occur between experimental and control group subjects on the post-test scores. However, this expectation is expressed as a null hypothesis, which states that "no difference" is expected. In a true experimental design, the investigator always states a null hypothesis that forms the basis for statistical testing. In research reports, however, usually the alternative hypothesis is stated that difference is expected.

There are three major characteristics of the true experiment.

*Randomization (R)
*Control
*Manipulation of an independent variable, or treatment (T)

Randomization is a powerful technique that increases control and reduces bias by neutralizing the effects of extraneous influences on the outcome of a study. For example, the threats to by historical events and maturation are theoretically eliminated. That is, based on probability theory such influences should affect subjects equally in both the experimental and control groups. Without randomization of subjects, you would not have a true experimental design.

The control group allows the investigator to see what the sample would be like without the influence of the experimental treatment. A control group theoretically performs the same at pretest and posttest occasions. It represents the characteristics of the experimental group before being changed by the experimental treatment. The control group is also a mechanism that allows the investigator to examine what has been referred to as either the Hawthorne effect or Halo effect. Without a control group, investigators would not be able to say that differences on posttest reflect that of experimental effect and not additional attention.

A. Randomized-Groups Design

Posttest-only designs conform to the norms of experimentation in that they contain the three elements of random assignment, control and treatment.

R	T	O_1	(Experimental group)
R		O_2	(Control group)

Theoretically, randomization should usually yield equivalent groups. However, the absence of the pretest makes it impossible to determine if random assignment successfully achieved equivalence between the experimental and control groups on the major dependent variables of the study. This design is most valuable when pretesting is not possible, but the research purpose is to seek casual relationship.

Furthermore, this experimental design is frequently used in the basic research using animal models in nutrition, biology, medicine, etc. because biological values of young animals are very much homogeneous.

Significant differences between O_1 and O_2 are due to treatment T. An independent t test is used to analyze the difference between the two groups.

This design may be extended to any number of levels of an independent variable.

R	T_1	O_1	(Experimental goup 1)
R	T_2	O_2	(Experimental group 2)
R		O_3	(Control Group)

Here, three levels of the independent variables, where one is the control and T_1 and T_2 represent two levels of treatment. This design can be analyzed by simple ANOVA. For example, an investigator studies the effect of deferent exercise levels on bone density of young rats: T1 is heavy training, T2 is light training, and T3, the control group, is no training. At the end of the experiment, the investigator measures bone density of rats, and compares them among the groups. Again, this design, even though no initial bone density measurements involved, is very much meaningful because bone densities of young animals are homogeneous as compared to those of humans in the next example.

B. Pretest-Posttest Randomized -Groups Design

In this design the groups are randomly assigned, but both groups are given a pretest as well as a posttest.

R	O_1	T	O_2	(Experimental group)
R	O_3		O_4	(Control group)

The major purpose of this type of design is to determine the amount of change produced by the treatment. This design threatens the internal validity of testing, but the threat is controlled, as the comparison of O3 to O4 in the control group includes the testing effect as well as the comparison of O1 to O2 in the experimental group

For instance, an investigator is investigating the effect of exercise levels on bone density in postmenopausal women. She has to measure bone density at the beginning of the study, and at the end of the study after treatment. Thus, she should compare the amount of change in bone density due to different exercise levels. Without premeaurement the experiment is not much meaningful because bone densities of postmenopausal women are not homogeneous because they have been affected by various factors throughout their lifetimes.

There are at least three common ways to do a statistical analysis of this design. First, a factorial repeated measures ANOVA can be used. One factor is the treatment versus no treatment, whereas the second factor is pretest versus posttest. A second analysis is to use simple ANCOVA with the pretest for each group (O1 and O3) used to adjust the posttest (O2 and O4). Sometimes the design is extended in other ways. For example, we could take the design in example of 2x3 factorial design, pretest, posttest, and treatment.

C. Factorial Design

Factorial designs offer more opportunity for multiple comparisons and complexity in analysis. In these designs the investigator evaluates the effects of either two or more independent variables (A and B).

		IV2	
		B1	B2
IV1	A1	A1B1	A1B2
	A2	A2B1	A2B2
	A3	A3B1	A3B2
R	A1B1	O1	
R	A1B2	O2	
R	A2B1	O3	
R	A2B2	O4	
R	A3B1	O5	
R	A3B2	O6	

Independent variable 1 (IV1) has three levels (A1, A2, A3), and IV2 has two levels (B1 and B2). This results in six cells (A1B1, A1B2, A2B1, A2B2, A3B1, A3B3) to which subjects are randomly assigned. At the end of the treatments, each cell is tested on the dependent variable (O1, O2, O3, O4, O5, O6). This design is analyzed by a 3x2 factorial ANOVA that the effects of IV1 (Fa), IV2 (Fb), and their interaction (Fab).

For instance, an investigator was interested in dietary fat sources and fiber sources on plasma cholesterol levels. He used fish oil, corn oil, and olive oil as a fat source, and oat bran and wheat bran as a fiber source. Rabbits were randomly assigned to six experimental groups, and fed their respective diet for six weeks. At the end of the study, the animals were sacrificed, and blood was drawn, and analyzed cholesterol levels. The data were analyzed by 2x3 factorial ANOVA, and tested the two main factors, fat source (Fa) and fiber source (Fb), and an interaction (Fab).

This design may also be extended to many other factorial designs such as 2x2x2, 2x3x4, and so on.

D. Solomon Four-Group Design

The Solomon four-group design represents a more complex experimental structure. It combines the true experiment and posttest-only design into one design structure. The strength of this design is that it eliminates the potential influence of the test-retest learning phenomenon by adding the posttest-only two group design to the true experimental design. The design is depicted as follows:

R	O1	T	O2	(Group1)
R	O3		O4	(Group 2)
R		T	O5	(Group 3)
R			O6	(Group 4)

The purpose is explicitly to determine whether the pretest results in increased sensitivity of the subjects to the treatment. This design allows a replication of the treatment effect (is O2>O4; O5>O6), an assessment of the amount of change due to the treatment (is O2-O1>O4-O3), an evaluation of the testing effect (is O4>O6), and an assessment of whether the pretest interacts with the treatment (is O2>O5). Thus, this is a very powerful experimental design. But the cost and time consuming are tremendous.

Example 11.1

Let us consider an example to illustrate the power of the Solomon four-group design and the nature of interaction effects. Suppose you wanted to assess **the effects of a Nutrition Education program (independent variable) on the eating disorder behaviors (anorexia nervosa, bulimia nervosa, binge eating disorder) of adolescents (dependent variable).** You would pretest groups by asking them questions regarding their anorexic, binge eating or bulimic activity and level of knowledge regarding risks of malnutrition as well as imbalanced nutrition. Then you would expose one group to the experimental treatment that involves attending a peer-led educational forum. On posttesting you find that levels of knowledge increased and that risk behaviors decreased in subjects who received the experimental program. However, you would not be able to determine the effect of pretest itself on the outcome. By adding group 3 (experimental treatment without a pretest), you can determine if the change in scores is as strong as when the pretest is administered (group 1). If group 2 (control) and group 3 (experimental) show no change but experimental group does, this change is a consequence of an interaction effect of the pretest and the intervention. If there is a change in experimental groups 1 and 3 and some change in control group but none in control group 4, there may be a direct effect of the treatment plus an interaction effect. This design allows the investigator to determine the strength of the treatment and the strength of the effect of pretesting on outcomes. If the groups that have been pretested show a testing effect, statistical procedures can be used to correct for it if necessary.

Experimental Groups

	Pre-test	Education	Post-test	Experimental Groups
R	Yes	Yes	Yes	Group 1 (Exp* with pre-test)
R	Yes	No	Yes	Group 2 (Con** with pre-test)
R	No	Yes	Yes	Group 3 (Exp without pre-test)
R	No	No	Yes	Group 4 (Con without pre-test)

*Experimental group; **Control group

3. Quasi-Experimental Designs

Quasi-experimental research designs have a very important role in research, especially in areas where the use of human subjects makes a strict experimental design impossible, unethical, or impractical. As the term "quasi-experimental designs" implies, these designs appear much like a true experiment. The exception would be in the lack of control compared to what one would normally expect to find in the experimental design.

The purpose of quasi-experimental design is to fit the design to settings more like the real world while still controlling as many of the threats to internal validity as possible. Quasi-experimental designs are characterized by the presence of control and manipulation, but do not contain random group assignment. In the time series and single-subject designs, although there is no control group, control is exercised through multiple observations of the same phenomenon both before and after the introduction of the experimental treatment. In non-equivalent group designs, the control is built in through the use of one or more comparison groups. Quasi-experiment is most valuable when investigator is attempting to look for change over time or a comparison between groups and the realities of the health or human service environment are such that random assignment is not feasible.

A. Ex Post Facto Design

In ex post facto research, variations in the variable of interest have not been manipulated by the researcher but rather have occurred at the various times prior to data collection in the natural course of events. Subjects have not been assigned randomly into treatment groups, and no control or comparison group has been studied.

Suppose a researcher is interested in investigating the cause of a high incidence of death associated with liver cancer in a certain province in China. This form of research will be ex post facto in nature because the researcher has to investigate several causes, including life styles, eating habits and diets etc., and attempt to find a common link between the victims. The research can at best only arrive at a probable cause since the researcher cannot control the independent variables by randomization or manipulations. The subjects either are already dead due to cancer of the liver or they already have cancer of the liver.

In the well-known cigarette smoking-cancer research, the smoking habits of a larger number of people were studied. This large group was divided into those who had lung cancer – or who had died of it – and those who did not have it. The dependent variable was thus the presence or absence of cancer. The investigator probed the subjects' backgrounds to determine whether they smoked cigarettes, and if so, how many. Cigarette smoking was the independent variable. The investigator found that the incidence of lung cancer rose with the number of cigarettes smoked daily. He also found that the incidence was lower in the cases of light smokers and nonsmokers. He came to the conclusion that cigarette-smoking caused lung cancer. This conclusion may or may not be true. But the investigator cannot come to this conclusion, although he can say that there is a statistically significant relation between the variables.

The reason he cannot state a causal connection is because there are a number of other variables, any one of which, or any combination of which, may have caused lung cancer. And he has not controlled other possible independent variables. He could say that cigarette smoking may be related to lung cancer.

Therefore, Ex post facto research is weaker than experimental design because of:

1. Inability to randomize
2. Inability to manipulate independent variables because of its retrospective nature
3. The higher possibility of incorrect interpretation, thus being misleading compared to experimental design.

However, many of these studies have linked life-style-related factors to various diseases and illnesses, e.g., smoking and lung cancer, drug- and alcohol-related congenital birth defects, and acquired immune deficiency syndrome.

Quasi-experiments will be further discussed in the following chapter.

Summary

Experimental research attempts to establish cause-and-effect relationships of independent and dependent variables. Cause and effect can be established only by the application of logical thinking to well-designed experiments. This logical process establishes that no other reasonable explanation exists for the changes in the dependent variable except the manipulation done as the independent variable.

Two concepts, external and internal validity, are vital to a conceptual grasp of all research design. Internal validity is the extent to which the research condition is controlled so that the independent variable causes an effect or change in the dependent variable. The stronger the control over an experimental condition, the higher degree of internal validity. Threats to internal validity include history, maturation, testing, instrumentation, regression to the mean, selection bias, experimental mortality, selection-maturation interaction, placebo effect, halo effect, Hawthorne effect, and Avis effect.

External validity is the ability to generalize the results to other subjects and to other settings. Four threats exist to external validity: reactive or interactive effects of testing, interaction of selection bias and experimental treatment, reactive effects of experimental arrangements, and multiple treatment interference.

Having high degrees of both internal and external validity is nearly impossible. The rigid controls needed for internal validity make it difficult to generalize the results to the real world, whereas studies with high external validity are usually weak in internal validity. Random selection of subjects and random assignment to treatments are the most powerful means of controlling most threats to internal and external validity.

The pre-experimental designs are weaker than the true experimental designs in terms of control. True experimental designs refer designs in which subjects are randomly selected and randomly assigned to either an experimental or control group. It includes randomized-groups design, pretest-posttest randomized-groups design, factorial design, and Solomon four-group design.

Quasi-experimental designs are characterized by the presence of control and manipulation, but do not contain random group assignment. However, these research designs have a very important role in research, especially in areas where the use of human subjects makes a strict experimental design impossible, unethical, or impractical.

216

References

Bracht G., Glass G.V. The external validity of experiments. Amer Educ Res J 1968; 5:437-474

Campbell, D.T., Stanley J.C. Experimental and Quasi-Experimental Designs for Research. Chicago, IL: Rand McNally, 1963.

Carroll J. Neglected areas in educational research. Phi Delta, Kappan 1961; 42: 339-343

Cook D. The Hawthorne effect in educational research. Phi Delta Kappan, 1962; 44:116-122

Cook, T.D., Campbell, D.T. Quasi-Experimentation: Design and Analysis Issues for Field Settings. Chicago, IL: Rand-McNally, 1979.

DeVellis, R.F. Scale Development: Theory and Applications. Newbury Park, CA: Sage, 1991.

Horn P.L., Lowell D.M., LiVolsi V.A., Boyle C.A. Reproducibility of cytologic diagnosis of human papilloma virus infection. Acta Cytologica 1985; 29: 692-694

Kerlinger, F. N. Foundation of Behavioral Research. New York: Holt, Rinehart & Winston, 1964.

Kirk, J., Miller, M. Reliability and Validity in Qualitative Research. Beverly Hills, CA: Sage, 1986.

Kirk, R.E. Experimental Design: Procedures for Behavioral Sciences. Monterey, CA: Brooks/Cole, 1982.

Kleinbaum, D.G., Kupper, L.L., Morgenstern, H. Epidemiologic Research: Principles and Quantitative Methods. Belmont, CA: Lifetime Learning Publications, 1982.

Lilienfeld, D.E., Stolley, P.D. Foundations of Epidemiology. (3rd ed.) New York: Oxford University Press, 1994.

Lipsey, M.W. Design Sensitivity: Statistical Power for Experimental Research. Thousand Oaks, CA: Sage, 1990.

Marshall, C., Rossman, G. Designing Qualitative Research. Newbury Park, CA: Sage, 1989.

Merton, R. Social Theory and Social Structure. New York: Free Press, 1949.

Miller, D. Handbook of Research Design and Social Measurement. Newbury Park, CA: Sage, 1991.

Morrow, J.R. "Generalizability Theory." In Measurement Concepts in Physical Education and Exercise Science. Safrit, M.J., Wood, T.M. (eds.), Champaign, IL: Human Kinetics, 1989.

Okolo, E.N. Health Research Design and Methodology. Boca, FL: CRC Press, 1990.

Pocock, S.J. Clinical Trials: A Practical Approach. Chichester, New York: John Wiley & Sons, 1983.

Silverman, W.A. Human Experimentation: A Guided Step into the Unknown. Oxford, England: Oxford University Press, 1985.

Stanly, J. "Studying Status vs. Manipulating Variables." In Research Design and Analysis. Bloomington, IN: Collier, R., Elam, S. eds. Phi Delta Kappa, 1961.

Steckler, A., McLeroy, K., Goodman, R., McCormick, L., Bird, S. eds. IntegratingQuantitative and Qualitative Methods. Health Education Quarterly (special edition), 19 (1), 1992.

Thomas, J.R., Nelson, J.K. Research Methods in Physical Activity. (3rd ed.) Champaign, IL: Human Kinetics 1996.

Webb, E.J., Campbell, D.T., Schwartz, R.D., Sechrest, L. Unobtrusive Measure: Nonreactive Research in the Social Sciences. Chicago, IL: Rand McNally, 1966.

Wechsler, D. The Measurement of Adult Intelligence. (4th ed.) New York: Psychological Corp, 1958.

Winer, B.J. Statistical Principles in Experimental Design. New York: McGraw-Hill, 1962.

===

CHAPTER 12

DESCRIPTIVE RESEARCH AND QUALITATIVE RESEARCH

Definition of Descriptive Study

Descriptive research is a study of status and is widely used in education, nutrition, epidemiology, and the behavioral sciences. Its value is based on the premise that problems can be solved and practices improved through observation, analysis, and description. The most common descriptive research method is the survey, which includes questionnaires, personal interviews, phone surveys, and normative surveys. Developmental research is also descriptive. Through cross-sectional and longitudinal studies, researchers investigate the interaction of diet (e.g., fat and its sources, fiber and its sources, etc.) and life styles (e.g., smoking, alcohol drinking, etc.) and of disease (e.g., cancer, coronary heart disease) development. Observational research and correlational studies constitute other forms of descriptive research. Correlational studies determine and analyze relationships between variables as well as generate predictions. Descriptive research generates data, both qualitative and quantitative, that define the state of nature at a point in time. This chapter discusses some characteristics and basic procedures of the various types of descriptive research.

Definition of Qualitative Research

Precise definitions for qualitative research are rarely found in the literature. The term qualitative research is an umbrella term referring to several research traditions and strategies that share certain commonalities. There is an emphasis on process, or how things happen, and a focus on attitude, beliefs, and thoughts – how people make sense of their experiences as they interpret their world. Qualitative research emphasizes inductive reasoning, whereby the researcher seeks to develop hypotheses from observations. The researcher is the primary research instrument, and the researcher's insight is the key instrument for analysis. Qualitative research has been an integral part of cross-cultural comparisons and descriptions of food habits in the nutrition and anthropology literature.

Descriptive Research --- Survey Research

Survey research is the most common type of descriptive research in dietetic, nutrition, and health areas. Survey research involves asking questions of a sample of individuals who are representative of the group or groups being studied. Such investigation may have a variety of purposes such as describing, comparing, correlating. Nutrition and food consumption surveys are used to assess the nutritional status, and dietary intake of population. Epidemiologic survey seeks to describe the distribution of disease and explain association between causative or associated factors and disease.

Surveys typically use four general procedures for data collection - the questionnaires, the personal interview, the Delphi method, and the normative survey. The essential task of any survey is to obtain information from a sample of respondents that relates to the questions being studied. Although this sounds simple enough, it can often be a formidable undertaking. Great care must be exercised in planning a survey investigation to facilitate successful execution of the study.

The advantages of survey design are that the investigator can reach a large number of respondents with relatively minimal expenditure, numerous variables can be measured by a single instrument. Disadvantages may include how the action processes of the survey design are structured. For example, the use of mailed questionnaires may yield a low response rate, compromising the external validity of the design. Face-to-face interviews are time consuming and may pose reliability problems.

1. Questionnaires

Questionnaires are written instruments and may be administered face to face, by proxy, or through mail. The development of questionnaire in survey research should always include a plan for testing its reliability and some level of validity to assure a level of accuracy and rigor to the study findings (see Chapter 10). In addition to the issues of reliability and validity, two other rather practical considerations influence of the selection of a data collection technique. First, the nature of sample determines the appropriate type of questionnaire for the survey. For instance, if you are interest in determining the accuracy of self-report of elders, you should use closed-ended questionnaires rather than open-ended questionnaires. Second, the nature of sample also determines the appropriateness of the data collection procedure. Such factors as level of education, socioeconomic background, verbal ability and cognitive status influence the selection of a data collection method. For example, college students are very familiar with a fixed-format type of self-administered questionnaire, but this format may not be appropriate with the present cohort of elderly.

If a questionnaire study is being designed, there is an obvious need to construct the instrument itself. The questionnaire must be constructed in such a manner that it will extract accurate information from the subjects. As a minimum, this means that questions must be written clearly and in a fashion that minimizes the possibility of misinterpretation by respondents.

The questionnaire represents the link between the researcher and data. It must, in large part, stand on its own since the researcher is not usually present to prompt a response or clarify areas where the subject may be confused. Since the researcher cannot personally work with each respondent, it is important to solve as many problems as possible as the questionnaire is being constructed. Keep the question easy. The more work the researcher spends on developing questions, the more pleasant the experience for the respondent. In multiple-choice-type questions, use options that will account for most of the possibilities. It is quite frustrating to be unable to indicate the response in questionnaire because the answer is not listed among the options. For example:

What is your current marital status? (check one)

_____Never married
_____Married
_____Widowed
_____Divorced

If a respondent is currently separated, she would have trouble with the question. The question should provide all possible answers and enough space so that the respondent may record pertinent additional information. For example, with the following question a respondent who is separated can answer. Moreover, a subject who is co-habitating can also respond to the question. He/she could mark at other and specify as co-habitating. For example:

What is your current marital status? (check one)

_____ Never married

____ Married
____ Widowed
____ Divorced
____ Separated
____ Other, Specify _____

Another mistake to avoid is the overlapping of categories. Make sure that only one response is appropriate for each individual.

What is your current grade point average (GPA)?

_____Under 2.00
_____2.00 - 2.50
_____2.50 - 3.00
_____3.00 - 3.50
_____Over 3.50

If subject's current GPA is 3.00, he/she will not know exactly how to answer the questionnaire. The construction of questionnaires should be checked carefully and pretested before administration of questionnaire to study subjects.

Type of Questions

1. Open-ended questions

Thomas and Nelson stated, "Open-ended questions are category of question in questionnaires and interviews that allows the respondent considerable latitude to express feelings and expand on ideas." Open-ended questions should be limited for using in questionnaires because of reduction of response rate.

Example 12.1

What are you doing to lose weight?
or
Why are you trying to lose weight?

However, sometimes the open-ended questions benefit over the closed-ended questions. For example, 24-hour dietary recall and food record methods are completed ended so that they can accommodate any level of food description detail necessary for addressing the questions of research interest. They can also accommodate any extent of diversity in the study population including various ethnic groups.

2. Closed-ended questions

There are a variety of forms: rankings, scaled items, and categorical responses.

Rankings

Subjects are asked to place responses in a rank order according to some criterion.

Example 12.2

Rank the following four drinks with regard to flavor. Use number 1-4, with 1 being the most preferred and 4 the least preferred.

____ Coke
____ Pepsi
____ Dr. Pepper
____ Seven-up

Scaled items

Subjects are asked to indicate the strength of their agreement or disagreement with some statement or the relative frequency of some behavior.

Example 12.3

Any of the following health symptoms are frequently the basis for medical attention. Circle the number indicating how often you have experienced each of the following:

1 = Practically never   2 = Infrequently   3 = Sometimes   4 = Fairly often   5 = Very often

Abdominal pain: 1 2   3 4   5

Low back pain : 1 2 3 4 5
Leg Pain   : 1 2 3 4 5

Categorical response

The categorical response question offers the subject responses, such as "yes" or "no." Answers to this type of questions that will be computer analyzed need to be numerically coded. Subjects should be asked to check the appropriate item or circle the appropriate number.

Example 12.4

1. Are you currently taking prescription hormones?

    1. Yes _____ 2. No _____
       If yes: 1. Estrogen? _____ 2. Progesterone? _____ 3. Both? _____

2. How do you take your hormones?

    1. By mouth _____ 2. Injection _____ 3. Patch _____ 4. Cream _____
    5. Other_____

All potential closed-ended responses should be precoded. For example, the following item from an interview about vitamin C has a series of precoded responses in the right-hand margin that will allow for rapid recording of data and rapid data entry:

If you take vitamin C: How many milligrams per vitamin C tablet?

    100 mg............................ 1
    250 mg............................ 2
    500 mg............................ 3
    1000 mg.......................... 4
    Don't know....................... 5

Researchers usually use various forms of questions together in one questionnaire form.

Order of Questions

Once questions and response categories have been formulated and formatted, the process of question placement or question ordering is followed. Some suggestions for question placement are summarized as follows (Sherry) :

1. Nonthreatening, simple-to-respond-to questions should come first.
2. Sensitive or difficult questions should be placed to the end of the questionnaire. Some researchers recommend that demographic questions also be placed at the end of a questionnaire. This information is uninteresting to the respondent and requires the repetition of fact that have been given many times.
3. A logical content order to question sequence should be established.
4. Within content categories, broad questions should start, and more specific questions should follow.
5. Transitional phases or visual distinctions should be provided between content sections.
6. Important items should not be placed last on the questionnaire, as they may be overlooked or left unanswered.
7. The question numbering system must be clear, not confusing.

Design Constructions of Questionnaire

The following are the summarized suggestions (Sherry):

1. Instruction should be concise.
2. The areas where answers are to be marked should be as close to the questions as possible.
3. The term questionnaire should not appear on the form itself. The term may have a bad connotation if a respondent has previously received poorly designed questionnaires.
4. The word over should be included at the bottom of the front side of a two-sided form.
5. Too many questions should not be placed on one page, creating a crowded appearance.
6. If the questionnaire is to be mailed, a name and return address should be printed directly on the questionnaire, even when a self-addressed return envelop is included, since the envelope may be misplaced or lost.
7. Colored paper can increase response rate.
8. Questions should be printed only on the front of pages.

9. Sufficient instructions should be provided to the interviewer to allow him or her to proceed in the desired question sequence.

10. Enough space should be allowed on the form so that the interviewer may record pertinent additional information.

2. Personal Interview

The steps for the interview and the questionnaire are basically the same. The focus here is only on the differences. The most obvious difference between the questionnaire and the interview is in the gathering of the data. In this respect, the interview is more valid because the responses are apt to be more reliable. Also, there are a much greater percentage of returns. However, interviews have disadvantage of being more costly and time consuming, and of being influenced by the relationship between the interviewer and the respondent. Generally, the interview uses smaller samples, especially when a graduate student is doing the survey.

Here is an example of methodology how the investigators improved the validity of nutrition survey through the personal interview over the questionnaire investigation (Koh & Caples)

The study was conducted from June through August 1974. The daily dietary intake data were collected on daily home visits by trained (40 hours) interviewers who were teachers and senior students majoring food and nutrition and general home economics at Alcorn State University. The students had taken several food and/or nutrition courses and understood well that the appropriate personal characteristics of an interviewer and the ability to establish rapport with the respondents were necessary to obtain accurate information. The interviewers were acquainted with the customs, the food habits, and the general life styles of these black families.

The 24-hr dietary recall method was used for seven consecutive days, including a weekend. The detailed nutritional history information that was described by Beal (1967) was modified on a daily base, including breakfast, lunch, dinner, and snacks. Individual forms were used for collecting information on the dietary intake of each family member. For satisfactory practical results in obtaining the dietary intake data for individuals, the food common to the area were listed, and enough spaces were supplied for recording other foods. To refresh their memory, the respondents were asked to record the time for each meal and snack. Food models were also used as memory aids to help respondents in estimating amounts of food consumed. Subjects received forms and directions for measuring food and reporting recipes a few days before the initial interview.

For young children, the informant was the person responsible for the child's feeding, usually the mother or baby-sitter. For subjects, ages six through twelve years, both the parents and the child were usually interviewed.

Because of Summer vacation, schools were closed during the interviewing period. Therefore, information about the availability and participation in the school lunch program was not considered. However, a continuous dietary intake study, including school lunches, was conducted for adolescents, ages eleven to eighteen years, in September 1974.

Standard food composition food tables were used for calculating nutrient intake. Home-made mixed foods or baked foods were assigned individual ingredients and serving sizes for each family member were estimated. The computer output was..."

In this study using personal interview methodology significantly increased the validity of survey as compared to that of questionnaire although the cost was high.

Personal interview can also be used where the respondent cannot read and write. It is also used for children and individuals with certain types of handicap, e.g., the blind. During a personal interview the interviewer reads out the instructions and then proceeds with the questions in the questionnaire. The interviewer reads the questions, followed by the responses, and the respondent chooses the best response applicable to him. This procedure is continued until all questions are answered.

Bias can be introduced when an interviewer influences the responses given by a respondent. For example, it will happen: 1) when the interviewer gives the responses by telling the respondent what to answer; 2) by making it difficult for the respondent to tell the truth by verbal or nonverbal expression; 3) arguing with the respondent about his/her response; 4) indicating the type of result you expect or want; 5) discussing the responses a respondent should give, etc.

Biasing a questionnaire is a serious matter because it makes the true results of the survey difficult to see. This is why it is very important that an interviewer should not give the responses or even suggest in any way the responses that a respondent should choose.

3. The Delphi Method

The Delphi survey method uses questionnaires but in a different manner than the typical survey. The procedures include the selection of the experts, or the informed persons who are to respond to the series of questionnaires. A set of statements or questions is prepared for consideration. The respondents are asked for their opinions on various issues, goals, and so on. The questionnaire is then revised as a result of the first round and sent to the respondents, asking them to reconsider their answers in light of the analysis of all respondents to the first questionnaires. Subsequent rounds are carried out, and the expert are given summaries of previous results and asked to revise their responses if appropriate. Their consensus about the issue is finally achieved through the series of rounds of analysis and subsequent considered judgments. An example of a case in which the Delphi technique might be useful is in investigating the potential role of registered dietitians in the year 2050.

4. The Normative Survey

This method involves establishing norms for abilities, performances, beliefs, and attitudes. A cross-sectional approach is used in that samples of people of different ages, genders, and other classifications are selected and measured. The steps in the normative survey are generally the same as in the questionnaire, the difference being in the manner in which the data are collected.

The researcher selects the most appropriate tests to measure the desired performances or abilities. This question of test selection is important in any type of research. The standardization of testing procedures is essential for establishing norms.

Nutrition Descriptive Research --- Nutrition Survey

Nutrition survey (descriptive study) is important for assessing nutritional status of a particular population. For example, the Nationwide Food Consumption Survey (NFCS) and the National Health and Nutrition Examination Surveys (NHANES) have provided information on the food and nutrient intake of the U.S. population. Nutritional assessment procedures were first used in surveys designed to describe the nutritional status of populations on a national basis. In 1955, the International Committee on Nutrition for National Defense (ICNND) was organized to assist developing countries in assessing their nutritional status and in identifying problems of malnutrition and the ways in which they could be solved. The ICNND teams conducted medical nutrition surveys in twenty-four countries, and consisted largely of clinical nutritionists, dentists, biochemists, food technologists, and agricultural, public health, and sanitation specialist; sometimes pediatricians, dermatologists, and ophthalmologists were also involved. The ICNND produced a comprehensive manual in 1963 describing their methods and interpretive guidelines (ICNND, 1963), with the intention of standardizing both the methods used for the collection of nutrition survey data and interpretation of the results.

Since the 1940, diet has been increasingly recognized as a major determinant of health and disease during adult years. To identify the relationship of diet to coronary heart disease, cancer, diabetes, and other chronic illnesses, epidemiologists and clinicians are including dietary assessment as an integral component in the design of cohort and case-control studies.

Dietary Intake

The first stage of a nutritional inadequacy is identified by dietary assessment methods. Unfortunately, there is no single dietary method suitable for all food consumption surveys or epidemiologic, nutritional status, and clinical investigations. Dietary methods are often classified according to "group" or "individual" methods.

Group Dietary Data

1. Food Disappearance Data

In the United States, the Economic Research Service (ERS) of the USDA compiles annual supply and use data for major commodities disappearing into civilian consumption. The total available food is determined from the sum of the beginning stocks, production estimates, and imports. The amount utilized during the year is the residual after subtracting food exported, purchased for the military, fed to livestock, put to non-food use, and ending stores. The civilian per capita consumption is then calculated by dividing the available quantity of each food by the estimated total civilian population. These per capita intakes are not accurate because food fed to pets, food waste, losses in transport and storage, food preparation losses, or spoilage are not deducted.

2. Household Food Intakes

In contrast to food availability, a closer approximation of food intakes may be derived from household estimates. The USDA has conducted periodic national surveys on household food use since 1936. Briefly, the person responsible for food planning and preparation recalls quantitative information on the foods purchased and utilized during a one-week period. This person is contacted a week prior to the interview and asked to keep a record of purchases and menus during the next seven days. A recall is obtained, with the aid of a detailed food list, of all foods used and brought into the household during the past week. This include the food consumed, along with foods discarded as garbage and fed to the animals, and thus is most likely an overestimate of the actual intake.

Individual Dietary Methods

1. 24-hour Recall

This technique is the most widely used dietary method in population studies. Information on all foods and estimated amounts consumed during the past 24 hours is obtained, beginning either with the present and working backward. The information usually is collected in an interview with a dietitian or trained interviewer. In face-to-face interviews, visual aids, such as food models, geometric models, pictures, or household measuring utensils, are often employed to help the subject estimate quantities consumed.

Since 1965, the USDA has used the 24-hour recall, combined with two subsequent days of food records, to derive daily individual intakes in their nationwide food consumption surveys. Similarly, the National Center for Health Statistics, which conducted the NHANES, has used the 24-hour recall, as well as food frequencies of various food items, in their surveys of stratified national probability samples of population.

The major strength of the 24-hour recall is its efficiency for comparing groups of people. Because 24-hour dietary recall is completed open-ended, they can accommodate any extent of diversity in the study population including various ethnic groups. It is relatively quick and easy to administer. The major limitation of 24-hour recall relates to the daily variation in food selection. Due to the large intra-individual variability in food and nutrient intakes of most persons, especially in the industrialized societies, a single 24-hour recall is not appropriate for estimating the usual intakes of a particular subject. To minimize this variation, the 24-hour recall for two week days and one weekend day is used for most of the nutrition surveys.

2. Food Records or Diaries

Food records or diaries require subjects to measure or estimate and record currently all foods consumed over a special period, usually three to seven consecutive days or multiple periods within a year. Foods are weighed, measured, or estimated. The method requires good instructions, adequate demonstrations, and ideally some observations. Generally, persons who agree to participate are dedicated, highly motivated, literate subjects, and that are most likely not representative of the general population.

A major strength of food records is that they do not rely on memory. Although food records are not necessarily error free and have not been validated, investigators assume that they approximate the "truth." Thus, the validation is actually relative. Along with its strength, this method has serious limitations as well. First, the particular period may be atypical for the subject, possibly due to illness, party meals, business obligations, or travel. Second, as noted previously, those persons who agree to keep food records may not be representative of the general population.

3. Diet Histories

The dietary history method, developed by Burke in the 1940s, describes usual dietary habits and includes a detailed interview of usual daily intake and its variations plus a food frequency check list and structured 3-day food records to serve as cross-checks.

For research concerning the etiology of disease, such as cancer and heart disease, investigators seek information on the usual diet consumed during a considerable period of time. This has led to further development and use of diet histories for large population studies. To simplify data collection and analysis and to increase objectively, most diet histories today are based on lists of selected foods and groups of items with similar nutrient values, used interchangeably in the diet. In the past, food items were often selected to test particular hypotheses concerning diet and disease, such as vitamin A and beta-carotene with lung cancer.

During the 1960s Heady et al. proposed short dietary questionnaires in which only the frequencies of a few food items would be needed to predict the usual intakes of various nutrients. The questionnaires were easy to administer, but difficult to validate. Nichols and coworkers (1976) used a food frequency questionnaire in the Tecumseh Heart Study population and failed to find an association between intake of fat, sugar, or starch-containing foods and level of serum cholesterol. Furthermore, the instruments omitted several foods of potential importance for diet and disease studies. Following a few trials, the limitations were apparent, and the research was discontinued.

Before using a diet history in a research setting, the instrument should be pretested extensively among representative samples of the population. Initially, the dietitian or trained interviewer should be present to identify problems needing clarification. Pretesting should be done repeatedly after each revision of the instrument.

Two well-known semiquantitative food frequency questionnaires are the Block Health Habits and History questionnaire designed for the cancer study and the Willet questionnaire for the 1980 Nurses' Health Study (see Chapter 10).

Epidemiologic Descriptive Research

Epidemiology seeks to describe the distribution of disease and explain associations between causative or associated factors and disease. Epidemiologic descriptive research is the reporting of the natural course of events, with particular emphasis on persons, place, and time. Quantifying correlations between these variables is also a part of descriptive epidemiology. Descriptive studies generated hypotheses that can be tested by analytical research methods.

For instance, in the review article on the cardiovascular effects of n-3 fatty acids, provide useful example of how different research designs are used in the process of gaining an understanding of disease causation, prevention, and treatment. Kromann and Green used **survey data** to describe the difference between Danes and Eskimos in age-adjusted mortality from heart attacks. Bang et al. demonstrated, using dietary intake and food composition analysis, that the difference in the diets of the two groups was the composition of the fat. In the next phase, clinical trials evaluating the effect of n-3 fatty acids on triglyceride concentrations demonstrated a reduction of the substance.

Epidemiologic Approaches to Diet and Disease

The field of nutritional epidemiology has developed in the concept that aspects of diet may influence the occurrence of human disease. In the mid-eighteenth century, observations that fresh fruits and vegetables could cure scurvy led Lind (1753) to conduct one of the earliest controlled clinical trials using lemons and oranges for the intervention group. Similarly, Goldberger used epidemiologic methods to determine that pellagra was a disease of nutritional deficiency, primarily associated with a corn meal subsistence diet in the southern United States.

Let us explain various epidemiologic approaches using cancer and diet relationship. Cancer is similar to other chronic disease, such as coronary heart disease, and to infectious and deficiency diseases because the chief underlying causes are environmental. Among these causes, food and nutrition are important modifiers of cancer risk.

1. Correlation Studies

Until recently, epidemiologic investigations of diet and disease consisted largely of ecologic or correlation studies, that is, comparisons of disease rates in populations with the population per capita consumption of specific dietary factors. Usually the dietary information in such studies is based on disappearance data, meaning the national figures for food produced and imported minus the food that is exported, fed to animals, or otherwise not available for humans. Many of the correlations based on such information are remarkably strong; for example, the correlation between meat intake and incidence of colon cancer is 0.85 for men and 0.89 for women in 23 countries (Figure 12-1; Amstrong and Doll, 1975; reproduced with permission).

Figure 12-1

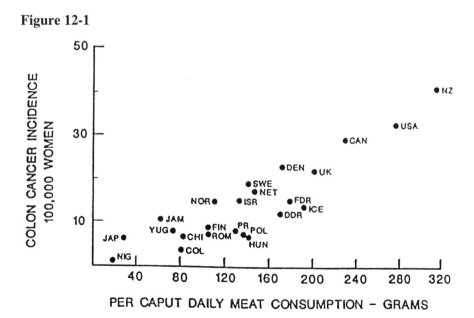

An obvious limitation of most correlation studies is that, while they may suggest a relationship between a specific environmental factor (such as an aspect of diet) and disease, the actual causal relationship may be with a different, diet-associated, confounding factor. For instance, diets high in meat (of the type eaten in industrialized societies) are associated with an increased risk of some cancers common in such societies. But diets high in meat are also likely to be high in fat, and may also be high in sugar and alcohol; in addition, people who eat such diets are relatively likely to have high incomes, live in urban areas and lead sedentary lives.

2. Special Exposure Groups

Groups with a population that consume unusual diets provide an additional opportunity to learn about the relation of dietary factors and disease. These groups are often defined by religious or ethnic characters and provide many of the same strengths as ecologic studies. In addition, the special populations often live in the same general environment as the comparison groups, which may somewhat reduce the number of alternative explanations for any differences that might be observed. For example, the observation that colon cancer mortality in the largely vegetarian Seventh-day Adventists is only about half that expected has been used to support the hypothesis that meat consumption is a cause of colon cancer.

Findings based on special exposure groups are subject to some of the same limitations as correlation studies. Many factors, both dietary and non-dietary, are likely to distinguish these special groups from the comparison population. Thus, another possible explanation for the lower colon cancer incidence and mortality among Seven-day Adventist population is that differences in rates are attributable to a lower intake of alcohol or to a higher consumption of vegetables.

3. Migrant Studies and Secular Trends

Migrant studies have been particularly useful in addressing the possibility that the correlations observed in the ecologic studies are due to genetic factors. For most cancers, populations migrating from an area with its own patterns of cancer incidence rates acquire rates characteristic of their new location, although, for a few tumor sites, this change occurs only in later generations. Therefore, genetic factors cannot be primarily responsible for the large differences in cancer rates among these countries. A few examples are shown in Figures 12-2 and 12-3.

4. Case-Control Studies

In case-control studies, patients with a specific type of disease (the case) and a comparable group of people without disease (the control) are identified from the same known source population. Information is then obtained from case and control subject, often by interview, about their earlier diet. It may then become evident that the cancer patients report higher levels of consumption of alcohol than the control subjects. Many of the weaknesses of correlational studies are potentially avoidable in case-control studies.

Case-control studies have some weaknesses of their own. Individuals may misreport their habitual past diets; if cases and controls differ in the accuracy of their dietary recall, the ensuing comparison will be biased. Also, some individual aspects of diet, especially nutrient content, may not vary greatly within a population, so case-control studies may not show wide ranges of cancer risk within that population.

5. Cohort Studies

In prospective cohort studies, the diets of a large group of healthy individuals are assessed and the group is then followed over time, usually for a decade or more, during which time a number of cohort members will develop cancer. The relationship of those cancers to specific characteristics of individual diets is then analyzed.

Prospective cohort studies avoid most of the methodological problems of other epidemiological studies. They also provide the opportunity to obtain repeated assessments of diet at regular intervals, which improves the validity individual dietary assessment and enables examination of the effects of diet on various cancers and on diseases other than cancer

The main limitations of prospective studies are practical. Even for common diseases, such as colon or breast cancer in industrialized societies, it necessary to enroll tens of thousands of subjects to have reasonable statistical power to determine relative risk. Hence, such studies are expensive. They are, therefore, of value mainly as means of investigating diseases common in economically developed countries. They are not a practicable means of investigating relatively uncommon diseases.

Cohort studies also usually depend on the use of self-administered food frequency questionnaires. Limitations on the validity of questionnaire assessment of usual diet are essentially same as for case-control studies.

236

6. Controlled Trials

Controlled trials use a control group of people given an inactive substance, and an intervention group given dietary constituents that may affect cancer risk. This is as close as epidemiology gets to the experimental designs used in animal studies.

For obvious ethical reasons, the study of factors believed to increase the risk of disease in human cannot usually be studied experimentally by deliberately exposing groups of people to the risk factor.

The preferred experimental method is randomized controlled trial in which people are assigned to an intervention or control group at random. In non-blind studies of foods, the control group may adopt the dietary behavior of the

Figure 12.2: Cancer incidence for selected cancers among Chinese men in different countries, 1983-87

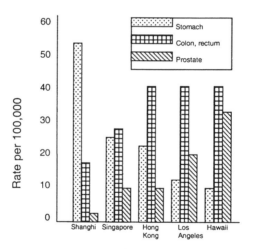

Age-adjusted to the World Standard Population
(From Parkin et al, 1992)

Figure 12.3: Cancer incidence for selected cancers in Japanese women by generation in Hawaii and Japan, 1968-77

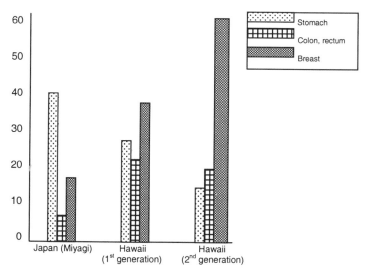

Age-adjusted to the World Standard Population
(From Kolonel et al, 1980)

treatment group if the treatment diet is thought by them to be beneficial. Such trends, which occurred in the US Multiple Risk Factor Intervention Trial of coronary disease prevention (Multiple Risk Factor Intervention Trial Research Group) may obscure a real benefit of the treatment.

7. Cross-Sectional Study

The structure of a cross-sectional study is similar to that of a cohort study except that all the measurements are made at once, with no follow-up period. Cross-sectional designs are very well suited to the goal of describing variables and their distribution patterns; in the Health and Nutritional Examination Survey (HANES), for example, a sample carefully selected to represent the U.S. population was interviewed and examined. The HANES is a major source of information about the health and habits of population. The survey provided estimates of such things as the prevalence of hypertension and the average daily fat intake.

The major strength of cross-sectional studies over cohort studies is that there is no waiting to see who will get the disease. This makes them relatively fast and inexpensive. A weakness of cross-sectional studies is the difficulty of establishing causal relationships from data collected in a cross-sectional frame.

8. Meta-Analysis and Pooled Analysis

The techniques of meta analysis and of pooled analysis have been developed to provide summaries of selected collections of studies. Meta-analysis and pooled analysis may select only prospective studies, or other studies judged to be of relatively high quality. Both have obvious strengths but, unless they include all relevant studies from a systematic literature review using comparatively objective criteria for selection, they can repeat or even magnify the bias of individual studies. A further potential problem is whether data from different studies are fully comparable: for example, different studies may use different quantified cut-off points for high or low intake or for quartiles or quintiles of intake.

Examples of Epidemiologic Descriptive Study in Diet, Nutrition and Diseases

The most influential descriptive study on diet and coronary heart disease (CHD) was the work of Keys (1980) relating the mean intake of dietary factors of 16 defined populations in seven countries to the incidence of heart disease in those same groups. As shown in Figure 12-4, intake of saturated fat as a percentage of calories was strongly correlated with coronary death rates ($r = 0.84$). In Keys' sample, the countries with low saturated fat intake and low incidence of CHD were less industrialized and were likely to have differed in many ways from the wealthier countries, in particular in physical activity, obesity, and at that time, smoking habits.

FIGURE 12-4: Ten-year coronary death rates of the cohorts plotted against the percentage of dietary calories supplied by saturated fatty acids (From Keys, 1980, reproduced with permission).

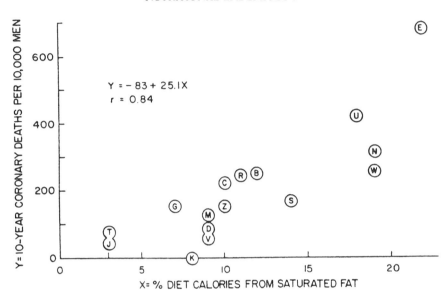

The descriptive epidemiology of cancer includes data on incidence, mortality and risk. The most basic information about cancer comes from statistics on cancer incidence and mortality. Cancer incidence rates are usually specified by sex and age. Cancer mortality rates are more widely available than cancer incidence rates, as the former can be derived from routinely collected national data on causes of death.

Qualitative Research

The term qualitative research is an umbrella term referring to several research traditions and strategies that share certain commonalities.

The researcher is the primary research instrument, and the researcher's insight is the key instrument for analysis. Qualitative strategies enable the researcher to record and understand people in their own terms. Research questions are not framed by delineating variables of testing hypotheses, but most often come from real-world observations. The data collected consist of detailed descriptions of people, events, situations, and conversations. Depth and detail are revealed through direct questions and careful descriptions of behavior. This material is usually supplemented by other data such as analysis of official documents, memos, records, photographs, and interviews with additional persons in order to cross-check and fill information gaps. Qualitative researchers collect data through sustained contact with people in their natural settings. They do not concisely intervene in any way or take action to change the situation under investigation. Data are analyzed inductively; the researcher builds concepts, explains processes, and develops hypotheses, rather than beginning with hypotheses and analyzing them deductively as in quantitative research (see Chapter 2).

The ethnographic cultural study is a good example of qualitative research. Everyone is intimately involved with his or her own culture; therefore, it is difficult for anyone to study another culture objectively. There are two different theoretical ways in which a culture may be viewed. The first and most common method is the etic approach, which is the viewpoint of an outsider to the culture. The other method is the emic approach, which is the viewpoint of an insider to the culture. Some anthropologists live in a cultural group for months or years in order to gain an insider's perspective.

For instance, nearly everyone must heard of classic ethnographic studies, such as the famous cultural research Margaret Mead conducted when she lived in the Samoan Islands in South Pacific. Mead interviewed the Samoans at length about their society, their traditions, and their beliefs.

The study of culturally based food habits is not an exact science; further, there are no absolute right or wrong ways to use food. It is sometimes difficult not to apply value judgments to other people's food habits, especially those that are repugnant within the context of one's own culture. The use of dog meat in some Asian cultures is an example. It is tempting to label such food habits as immoral or disgusting, yet this is cultural bias. Most Asians do not share Americans' adoration of dogs as pets. Instead, within the context of their cultures, dogs are considered and acceptable food source.

As the above example, qualitative research has always been an integral part of cross-cultural comparisons and descriptions of food habits in the nutrition and anthropology literature, but it has also become routine in food product development and sensory evaluation, food and nutrition marketing, and nutrition education materials development and evaluation. The unique assumptions underlying qualitative research, the various approaches that may be used, and types of information that they can produce are of importance to nutritionists.

Methods of Collecting Data

In qualitative studies, the primary instrument for gathering information is the investigator. As Fetterman describes: "The ethnographer is a human instrument. Relying on all its senses, thought, and feelings, the human instrument is a most sensitive and perceptive data gathering tool."

The most common sources of data collection in qualitative research are interviews, observations, and researcher-designed instruments

Interviews

The interview is the most common source of data in qualitative studies. The person-to-person format is most prevalent, but occasionally group interviews are conducted. The interviewer must be alert to both verbal and nonverbal messages and be flexible in rephrasing pursuing certain lines of questioning. The researcher must use words that are clear and meaningful to the respondent. The interviewer has to be a good listener.

Observation

The researcher usually spends extensive periods of time in the natural setting of the participants, observing the events, processes, and activities in which they are involved. Anthropologists who go to a new country, begin learning the language, and participating in the culture are an example.

Earlier studies relied on direct observation with note taking and coding of certain categories of behavior. More recently, videotaping has been the method of choice. Other recording devices include narrative field logs, and diaries, in which researchers record their reactions, concerns, and speculations.

Documents

The researcher is the main producer of field notes and transcripts. Personal documents written by respondents can also yield useful information. These materials can be especially valuable in clarifying what respondents have experienced.

Photographs and videos

Researchers use photographs to learn how people interpret their world. Photographs and videos taken in the field can be a way to help the researcher remember detail the situations. Showing a photograph to respondents can stimulate their insights and serve as a source of data.

Triangulation

This is the process of cross-validation among researchers, research methods, and data sources. Cross-checking across different methods such as observations, interviews, and physical evidence contributes greatly to the study validity.

Analyze the Data

Qualitative researchers begin analysis of data almost immediately, and it proceeds along with data collection. Thus, simultaneous data collection and analysis allow the researcher to work more effectively. Another difference between quantitative and qualitative data analysis is that qualitative data are generally presented through words, descriptions, and images, whereas quantitative analysis is typically presented through numbers.

Summary

Descriptive research encompasses many different techniques. The techniques are measures of status although some might seek opinions or even future projections in the responses. The most common descriptive research technique is the survey. The survey includes questionnaires and personal interview. Personal interviews usually yield more valid data because of the personal contact and the opportunity to make sure that respondents understand the questions.

However, interviews have disadvantage of being more costly and time consuming, and of being influenced by the relationship between the interviewer and the respondent.

The Delphi survey is frequently used to survey expert opinion in an effort to help make decisions about practices, needs, and goals. The normative survey is designed to obtain norms for abilities, performances, beliefs, and attitudes.

Nutrition survey is important for assessing nutritional status of population (s). Various dietary methods for nutrition survey are discussed. Epidemiology seeks to describe the distribution of disease and explain association between causative or associated factors (mainly, dietary and life style factors) and disease. Correlation studies, special exposure group, migrant studies, case-control and cohort studies are the important approaches for descriptive epidemiologic study.

The term **qualitative research** is an umbrella term referring to several research traditions and strategies that share certain commonalities. There is an emphasis on process, or how things happen, and a focus on attitudes, beliefs, and thoughts - how people make sense of their experiences as they interpret their world.

Qualitative research methods include field observations, interviews, case studies, ethnography, and narrative reports. Qualitative research has always been an integral part of cross-cultural comparisons and descriptions of food habits in the nutrition and anthropology literature, but it has also become routine in food product development and sensory evaluation, food and nutrition marketing, and nutrition education materials development and evaluation.

A major difference between quantitative and qualitative data analysis is that qualitative data are generally presented through words, descriptions, and images, whereas quantative analysis is typically presented through numbers.

244

References

Achterberg C. Qualitative methods in nutritional education evaluation research. J Nutr Educ 1988; 20: 244

Agar, M.H. The Professional Stranger: an Informal Introduction to Ethnography. Academic Press, New York: Academic Press 1980.

Amstrong B., Doll R. Environmental factors and cancer incidence and mortality in different countries, with special reference to dietary practices. Int J Cancer, 1975; 15:617-631

Bang H.O., Dyerberg J., Hjorne, N. The composition of food consumed by Greenland Eskimos. Acta Med Scand 1976; 200: 69-75

Bailey, K.D. Methods of Social Research. New York: Free Press, Macmillan Publishing Co, Inc, 1978.

Beal V.A. The nutritional history in longitudinal research. J Am Dietet A 1967; 51: 526-531

Berdie D.R. Questionnaire length and response rate. J Appl Psychol 1973; 58:278-280

Berdie, D.R., Anderson J.F., Niebuhr, M.A. Questionnaires: Design and Use. Metuchen, N.J.: Scarescrow Press, 1986.

Buell P. Changing incidence of breast cancer in Japanese-American women. JNCI 1973; 51:1479-1483

Dyberg J., Bang H.O., Hjorne N. Fatty acid composition of the plasma lipids in Greenland Eskimos. Am J Clin Nutr 1975; 28: 958-961

Frank G. Life history model of adaptation to disability: the case of a congenital amputee. Soc Sci Med 1984; 19: 639-645

Fetterman, D.L. "A Walk Through the Wilderness: Learning to Find Your Way." In Experiencing Fieldwork: An Inside View of Qualitative Research. Shaffir, W., Stebbins, R. eds. Newbury Park, CA: Sage, 1991.

Fieldhouse, P. Food & Nutrition: Customs & Culture. New York: Croom Helm, 1986

Fielding ,N.G., Fielding, J.L. Linking Data. Beverly Hills, CA: Sage, 1986.

Firestone W.A. Meaning in method: The rhetoric of quantitative and qualitative research. Educational Researcher 1987; 16:16-21

Geertz, C. The interpretation of Cultures. New York: Basic Books, 1973.

Glaser B.G., Strauss, A.L. The Discovery of Grounded Theory: Strategies for Qualitative Research. New York: Aldine Publishing Co, 1967.

Goldberger, J.E. Goldberger on Pellagra. Baton Rouge, LA: Louisiana State University Press, 1964.

Greer J.G. What do open-ended questions measure? Public Opinion Quart 1988; 52:365-371

Gordis, L. Epidemiology. Philadelphia PA: Saunders, 1996.

Hammersley, M., Atkinson, P. Ethnography: Principles in Practice. London: Tavistock, 1983.

Headland, T.N., Pike, K.L., Harris, M. Emics and Etics: The Insider/Outsider Debate. Newbury Park, CA: Sage, 1990.

Hodge, R., Kress, G. Social Semiotics. Ithaca, NY: Cornell University Press, 1988.

Holbrook J.T., Patterson K.Y., Bodner J.E., Douglas L.W., Veillon C., Kelsey J.L. Mertz W, Smith J.C. Sodium and potassium intake and balance in adults consuming self selected diets. Am J Clin Nutr 1984; 40: 786-793

Human Nutrition Information Service, US Department of Agriculture, Food Consumption: Households in the United States, Spring 1977, Washington, D.C.: Government Printing Office, Publication H-1, 1982.

ICNND (International Committee on Nutrition for National Defense) Manual for Nutrition Surveys. Second edition. Superintendent of Documents. Washington, D.C.: U.S. Government Printing Office, 1963

Jacob E. Qualitative research traditions: A review. Rev Educ Res 1987; 57: 1-4

Jelliffe, D.B. The Assessment of the Nutritional Status of the Community. WHO

246

Monograph 53. Geneva: World Health Organization, 1966.

Keys, A. Seven Countries: A Multivariate Analysis of Death and Corornary Heart Disease. Cambridge, MA: Harvard University Press, 1980.

Kittler, P.G., Sucher, K. Food and Culture in America. New York: Reinhold, 1989.

Kirk, J., Miller, M.L. Reliability and Validity in Qualitative Research. Beverly Hills, CA: Sage, 1986.

Koh E.T., Caples V. Nutrient intake of low-income, black families in Southwest Mississippi. J Am Dietet A 1979; 75:665-670

Kolonel, L.N., Hinds, M.W., Hankin, J.H. "Cancer Patterns Among Migrant and Native-Born Japanese in Hawaii in Relation to Smoking, Drinking, and Dietary Habits." In Genetic and Environmental Factors in Experimental and Human Cancer. Gelboin, H.V. et al. eds. Tokyo, Japan: Science Press, 1980.

Kromann N., Green A. Epidemiological studies in the Upernavik district, Greenland: Incidence of some chronic disease 1950-1974. Acta Med Scand 1980; 401-405

Krueger, R. Focus Groups: A Practical Guide for Applied Research. Newbury Park, CA: Sage, 1988.

Langness, L.L. Frank, G. Lives: An Anthropological Approach to Biography. Novato, CA: Chandler & Sharp, 1981.

Lind, J. (1753) A Treatise on the Scurvy. Reprinted Edinburgh: Edinburgh University Press, 1953.

Lock, L.F. "The Question of Quality in Qualitative Research." In Proceedings of the 5th Measurement and Evaluation Symposium. Nelson, J.K. ed. Baton Rouge, LA: Louisiana State University Press, 1987.

Lofland, J., Lofland, L. Analyzing Social Settings: A Guide to Qualitative Research. Belmont, CA: Wadsworth, 1984.

Marshall, C., Rossman, G.B. Designing Qualitative Research. Newbury Park, CA: Sage, 1989.

Mauser, J.S., Kramer, S. Epidemiology – An Introductory Text. Philadelphia PA: Saunders, 1985.

McClendon M.J., O'Brien D.J. Question-order effects on the determinants of subjective well-being. Public Opinion Quart 1988; 52:351-364
McCraken, G. The Long Interview. Newbury Park, CA: Sage, 1988.

McMichael A.J., McCall M.G., Hartshorne J.M., Woodings T.L. Patterns of gastro-intestinal cancer in European migrants to Australia: The role of dietary changes. In J Cancer 1980; 25:431-437

Miles, M.B., Huberman, A.M. Qualitative Data Analysis: A Sourcebook of New Methods. (2nd ed.) Thousand Oaks, CA: Sage, 1994.

Montgomery A.C., Crittenden K.S. Improving coding reliability for open-ended questions. Public Opinion Quart 1977; 41: 235-243

National Center for Health Statistics: Plan and Operation of the HANES. Vital and Health Statistics, Series 1, Nos 10a and 10b, DHEW Pub No (HSM) 73-130. Washington, D.C.: U.S. Government Printing Office, 1973.

O'Brien T., Dugdale V. Questionnaire administration by computer. J Market Res Soc, 1978; 20:228-237

Okolo, E.N. Health Research Design and Methodology. Boca Raton, FL: CRC Press 1990.

Parkin, D.M., Muir, C.S., Whelan, S.L., Gao, Y.T., Ferlay, J., Powell, J. Cancer Incidence in Five Continents VI IARC Sci Publ 120, Lyon: International Agency for Research on Cancer, 1992.

Peterkin B.B., Rizek R.L., Tippett K.S. Nationwide food consumption survey, 1987. Nutr Today 1988; 23:18-24

Poe G.S., Seeman I., McLaughlin J., Mehl E., Dietz M. Don't know boxes in factual questions in a mail questionnaire: Effects on level and quality of response. Public Opinion Quart 1988; 52: 212-222

Poikolainen K., Karkkainen P. Nature of questionnaire options affects estimates of alcohol intake. J Stud Alcohol 1985; 46: 219-222

Potter, J.D. Food, Nutrition and the Prevention of Cancer: A Global Perspective.

248

Washington, D.C.: American Institute for Cancer Research, 1997.

Rose, G.A., Blackburn, H. Cardiovascular Survey Methods. Geneva, Switzerland: World Health Organization 1968.

Runcie, J.F. Experiencing Social Sresearch. Homewood, IL: Dorsey Press, 1976.

Sandelowski, M. The Problem of Rigor in Qualitative Research. Adv Nurs Sci April 27-37, 1986.

Schutz R.W. Qualitative research: comments and controversies. Research Quarterly Exercise Sport,1989; 60:30-35

Sheatsley, P.B. "Questionnaire Construction and Item Writing." In Handbook of Survey Research. Rossi, P.H., Wright, J.D., Anderson, A.B. eds. Orlando, FL: Academic Press, 1983.

Sherry, B. "Epidemiologic Analytical Research." In Research. Monsen, E.R. Chicago, IL: American Dietetic Association, 1992

Spradley, J. The Ethnographic Interview. Chicago, IL: Holtz, Rinehart & Winston, 1979.

Spradley, J.P. Participant Observation. New York: Holt, Rinehart & Winston, 1980.

Sudman, S., Bradburn, N.M. Asking Questions: A Practical Guide to Questionnaire Design. (2nd ed.) Washington, D.C.: Jossey-Bass Publisher, 1983.

Taylor, S.J., Bogdan, R. Introduction to Qualitative Research Methods. (2nd ed.) New York: Wiley, 1984.

Thomas, J.R. Nelson, J.K. Research Methods in Physical Activity. (3rd ed.), Champaign, IL: Human Kinetics, 1996.

Van Maanen, J. ed. Qualitative Methodology. Newbury Park, CA: Sage, 1983.

Willett, W. Nutritional Epidemiology. (2nd ed.) New York: Oxford University Press, 1998.

PART IV

===

WRITING THE RESEARCH PROPOSAL AND RESULTS

Part IV explains how to present research results via oral and poster presentations, and how to write theses, dissertations, journal articles, and research proposals.

Chapter 13 helps you organize and write the results, discussion and abstract sections. This chapter also explains how to prepare tables, figures, photographs and illustrations and where to place them in the research report.

Chapter 14 demonstrates how to write theses, dissertations, and journal articles. We also discuss how to prepare and present oral and poster presentations.

Chapter 15 explains how to write a research proposal. We discuss the essential components of research proposals, and how to write a successful grant application.

CHAPTER 13

===

RESULTS, DISCUSSION, AND ABSTRACT

Introduction

The results section is the most important part of the research report. The reports generated through research are perhaps the most common medium for researcher to communicate with both academic audience and the practical reader in the world of men. In another word, this section presents your contribution to knowledge, that is, what you found. Journal articles, master's theses, and doctoral dissertations all have the same objectives.

A logical place is to state whether or not the hypothesis being tested was rejected. The best way may be to address each of the tested hypotheses: Describe the analysis and results for the first hypothesis first, then describe them for the second hypothesis, and so on. On other occasions, the results may be organized around the independent or dependent variables of interest. The results should be concise and effectively organized and should include appropriate tables, graphs, figures, and/or photographs.

Only the most important tables, figures, graphs, and/or photographs should be included in the results in journal articles. More data than journal articles can be presented in thesis and dissertation, and the remaining less important results may be placed in the appendix in thesis or dissertation. The primary or raw data can also be presented in the appendix in thesis or dissertation. Use graphs as an alternative to tables with many entries; do not duplicate data in graphs and tables. Regardless of the type of illustration used, it should contain enough information to be understandable without referring to the text. In the text, describe the main conclusions to be reached based on each table or figure and point out highlights that the reader may otherwise overlook.

Statistical figures and graphs (e.g., histograms) should be professionally drawn. Scientific journals specify a style for tables and require its use. However, many do not have rigid specifications for graphs and other figures.

Use of Tables in Data Presentation

Tables usually present research data in columns and rows to demonstrate relationships between independent variables (manipulation) and dependent variables (effects). They may consist of means (M) and standard deviations (SD), or standard error of the mean (SEM) for dependent variables under the important conditions including probability levels (e.g., $p<0.01$) at which group differences were significant. For categorical data, frequencies and percentages, chi-squares, etc. should be presented. These are basic descriptive data that allow other researchers to evaluate your findings.

Tips for Clear Tables

1. Every table should contain enough information to be understandable without referring to the text.
2. Every table should have a title describing the main information. The title is usually located above the table.
3. All columns and rows should be titled. The columns and rows should be arranged in a logical order to facilitate comparisons when relevant.
4. All measurements should have clear units.
5. All relevant information needed to interpret the table should either be incorporated into the table itself or included as footnotes on the same page.
6. Tables are usually located immediately after their introduction in the text.
7. Avoid vertical lines within the table or as borders. These lines often complicate the table and make it difficult to read.
8. Horizontal lines may be used to separate the title of the table from entries. Column total may also be separated by horizontal lines.

Example 13-1

Table with Mean ±SD and different superscripts explaining how groups means are different

---

EFFECTS OF SEX HORMONES AND DIETS ON FINAL BODY WEIGHTS, FEMUR WEIGHTS, BONE MINERAL CONTENT AND BONE MINERAL DENSITY OF ORCHIECTOMIZED RATS FED EITHER HIGH FRUCTOSE OR CORNSTARCH WITHOUT MAGNESIUM

Diets	Hormones	FBW[1] (g)	Femur Bone		
			Weights (g)	Bone Mineral Density (g/cm^2)	Bone Mineral Content (g)
Fructose	Orchiectomy (8)[2]	384 ± 53[3]	0.947 ± 0.06	0.598 ± 0.05	0.290 ± 0.01
	Sham (8)[4]	404 ± 20	0.930 ± 0.06	0.598 ± 0.04	0.291 ± 0.01
	Oestrogen (8)	320 ± 24	0.989 ± 0.05	0.661 ± 0.03	0.325 ± 0.01
	Testosterone (8)	338 ± 48	0.920 ± 0.05	0.588 ± 0.04	0.288 ± 0.01
Cornstarch	Orchiectomy (8)	368 ± 41	0.931 ± 0.04	0.578 ± 0.03	0.281 ± 0.01
	Sham (8)	355 ± 64	0.915 ± 0.06	0.578 ± 0.04	0.285 ± 0.01
	Oestrogen (8)	307 ± 25	0.955 ± 0.06	0.650 ± 0.04	0.315 ± 0.01
	Testosterone (8)	319 ± 46	0.870 ± 0.07	0.537 ± 0.04	0.270 ± 0.01
		Analysis of Variance			
Hormone Effects	Orchiectomy	376^{A5}	0.939^{AB}	0.588^{B}	0.286^{B}
	Sham	382^{A}	0.923^{B}	0.590^{B}	0.289^{B}
	Oestrogen	314^{B}	0.972^{A}	0.656^{A}	0.320^{A}
	Testosterone	329^{B}	0.895^{B}	0.564^{B}	0.279^{B}
Diet Effects	Fructose	361^{A}	0.947	0.611^{A}	0.299^{A}
	Cornstarch	338^{B}	0.917	0.588^{B}	0.288^{B}

---

[1]Final Body Weight [2]n = 8 [3]Mean ± SD [4]Sham-orchiectomy [5]Means with different superscripts are significantly different at P < 0.05 by Duncan's multiple range test

First, this table contains enough information so that readers can understand or explain the experimental results as follows without referring to the text:

This table demonstrates relationships between independent variables and dependent variables.

Independent variables:

Diets ………………….. Fructose or cornstarch without magnesium

Sex hormones…………Orchiectomy,sham, oestrogen or testosterone

Dependent variables:
    Final body weight, femur weight, bone mineral content, bone mineral density

The experimental design was 2x4 factorial design, having 2 diet groups and 4 hormone groups. Eight animals were in each group, and there is no loss of animals during experimental period. The data were analyzed by 2x4 factorial analysis of variance (ANOVA), and by Duncan's multiple range test as a follow-up test.

ANOVA table shows that final body weights were significantly affected by hormones and diets ($p < 0.05$). Although all animals were pair fed (16g/rat/day)(not shown in the Table), either oestrogen or testosterone injection decreased final body weights significantly. The animals fed fructose had greater final body weights than those fed cornstarch.

Oestrogen treatment increased femur weights significantly in the orchiectomized animals compared to sham-control animals, whereas testosterone treatment decreased femur weights. Bone mineral content and bone mineral density were greatest in the oestrogen treated group among all experimental groups. Bone mineral content and bone mineral density were also significantly affected by diets, showing greater bone mineral content and bone mineral density in the fructose- than in the starch-fed animals. In the oestrogen-treated animals, the fructose group had the greater bone mineral density than starch group ($325g/cm^2$ vs., $315g/cm^2$).

Let us examine the Table according to the remaining suggested tips. The title, "Effects of sex hormones and diets on final body weight, femur weight, bone mineral content and bone mineral density of orchiectomized rats fed either high fructose or cornstarch without magnesium," has fully described the main information and located above the Table. All columns and rows were titled, and arranged in a logical order to facilitate relevant comparisons. All measurements have units, g or g/cm^2, in the Table. All relevant information needed to interpret the table, sample size, means and SD, Duncan's multiple test, significant level, etc., were incorporated into the Table. The Table has no vertical lines at all, whereas it has a few horizontal lines to separate the title from entries and for total columns.

Use of Graphs in Data Presentation

A graph is a drawing or figure used to convey statistical information. Graphs greatly simplify communication because figures have easily visualized shapes.

Some general rules for constructing graphs are:

1. Every graph should contain enough information to be understandable without referring to the text.
2. Every graph should have a title. This is located next to the figure number and may be placed above or below the graph. The common method is to place the title of a graph beneath the figure.
3. All symbols used on the graph should be clearly identified.
4. All graphs, with the exception of pie graphs are plotted on a horizontal axis (called the X axis or abscissa) and a vertical axis (called the Y axis or ordinate). The horizontal axis usually represents dates, number of trials, or other fixed values, i.e., independent variables, whereas the vertical axis represents the corresponding frequencies, percentages, or measurements, i.e. dependent variables.

Standard graph types include line graphs, scatter graphs, histograms, frequency polygons, bar graphs, and pie charts.

The Line Graphs

The line graph consists of a series of straight lines connecting the points that correspond to the values on the X and Y axes. Research articles commonly use SEM values in graphs to show data variability. The error bars in the figure of Example 13-2 are the vertical lines connecting the mean point to the mean ± SEM points. The error bars may be adjusted to represent other quantities such as the mean ± a confidence interval or the mean ± standard deviation.

Example 13-2 Line graph

The line graphs with mean ± SEM and different superscripts explaining how group means are different (Yu and Morris, J Nutr, 1999) Used with permission

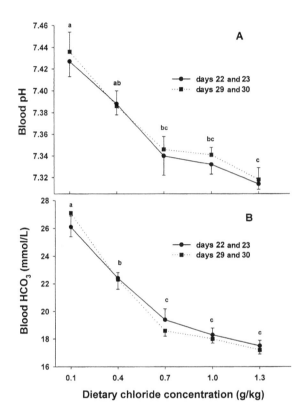

Blood pH (panel A) and HCO_3 (panel B) of kittens given a purified diet with various levels of chloride. Each point represents the mean of six kittens (three males and three females); vertical bars are SEM. There was a significant diet but no time effect on blood pH and HCO_3 when the data of d 22 through 30 were analyzed by ANOVA (diet $P < 0.01$, time $P > 0.05$). Points not sharing the same letter are significantly different ($P < 0.05$, Tukey's test for both blood pH and HCO_3 on d 29 and 30.)

The graph contains enough information so that readers can explain the results without referring to the text as follows:

The graph demonstrates relationship between independent variables and dependent variables.

Independent variables:
Dietary chloride concentrations, and duration of feeding (day 22 and 23 vs. days 29 and 30

Dependent variables:
Blood pH (Figure A), and blood bicarbonate (Figure B)

The experimental design was 2x5 factorial design having two groups of feeding duration and five chloride concentrations. Each point represents the mean of six kittens (three males and three females). The data were analyzed by 2x5 factorial analysis of variance (ANOVA), and followed by Tukey's test for both blood pH and bicarbonate on d 29 and 30.

There was a significant diet but not time effect on blood pH and bicarbonate when data of d 22 through 30 were analyzed by ANOVA (diet $p < 0.01$, time $p > 0.05$). Points not sharing the same letter are significantly different ($p < 0.05$).

As dietary Cl concentration decreased from 1.3 to 0.1 g/kg, there was evidenced by a significant elevation of blood pH (Figure A), and an increase in blood bicarbonate (Figure B). Both blood pH bicarbonate were similar at d 22-23 and d 29-30, indicating that the kittens had attained a stable state.

The Histogram

A histogram consists of a number of columns where the height of each column represents a corresponding frequency on the vertical axis. The higher the column, the greater the frequency is. However, it illustrates continuous data and compares the areas under the graph. One disadvantage of the histogram is the necessity of beginning the vertical scale with 0. (Example 13 -3)

Example 13-3 Histogram

Histograms illustrate continuous data and compare the areas under the graph.

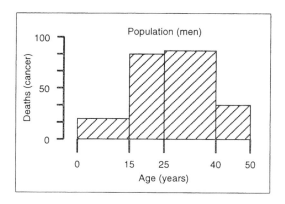

The Bar Graph

The bar graph is essentially the same as a histogram except that (1) horizontal bars replace columns, and (2) the categories are separate rather than continuous (Example 13-4). However, these are minor differences which may statisticians are willing to overlook since no more information is conveyed by a histogram than by a bar graph.

Example 13-4 Bar Graphs

Exogenous oestrogen affects calcium metabolism differently from exogenous testosterone in ovariectomized rats fed a high fructose diet severely deficient in magnesium (Koh et al.)

A. Effects of exogenous oestrogen or testosterone on serum PTH of 10-week old ovariectomized female and orchiectomized male rats fed magnesium-deficient fructose diets.

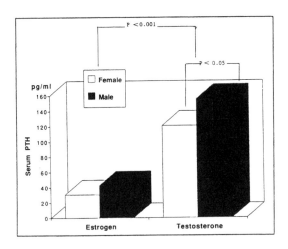

B. Effects of exogenous oestrogen or testosterone on urinary calcium of 10-week old ovariectomized female and orchiectomized male rats fed magnesium-deficient fructose diets.

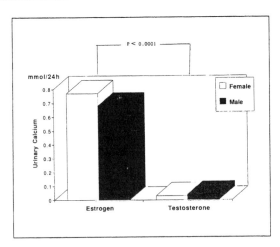

The graphs present relationships between the following independent variables and dependent variables:

Independent variables: Hormones....... Estrogen, testosterone
                        Sex................. Orchiectomized male, ovariectomized female
Dependent variables: Serum parathyroid hormone (PTH) (Figure A)
                      Urinary calcium (Figure B)

The experimental design was 2x2 factorial design, having two hormone groups, estrogen and testosterone, and both sexes, male and female animals. The animals were 10-week old ovariectomized female and orchiectomized male rats. They were treated with exogenous estrogen or testosterone.

The figure does not indicate statistical analyses, and either SD or SEM. But it shows significant differences among groups with p levels.

Circulating serum PTH was almost four times higher in testosterone-treated animals (Fig A). Circulating PTH was significantly higher in male testosterone-treated animals than in female testosterone-treated animals.

The effects of hormone treatment on urine calcium excretion of ovariectomized female and orchiectomized male rats are shown in Fig. B. Urinary calcium was 22 times higher in oestrogen-treated animals than in testosterone-treated animals. This difference was even greater in female rats than in male rats, the values being respectively 29 times in females vs. 17 times in males.

The Pie Chart

The pie chart is a circular figure that is especially useful for demonstrating how various proportions of a total are related to one another, such as for showing the proportion of males and females in a group. The data in a pie chart are typically arranged so that, as one proceeds clockwise (or counterclockwise), the percentage made up from each component steadily decreases (or increases). The figure in Example 13-5 is arranged as counterclockwise.

Example 13-5 Pie Chart

Presence of inulin and oligofructose in the diets of Americans (Moshfegh et al, 1999) (Used with permission)

See Figure 13-5: Contribution of food sources to inulin and oligofructose in American diets on the following page.

The figure was arranged as one proceeds counter clockwise, the percentage made up from each component steadily decreases.

The Figure A illustrates the contribution of food sources to inulin and Figure B, to oligofructose in American diets. When consumption of food was accounted for, wheat was the most important source of both components in American diets followed by onions. Wheat and onions were consumed by >90% of Americans on any given day in 1994-1996. Wheat contributed 69% of the inulin and 71% of the oligofructose. Onions contributed ~25% of each of these components.

The Scatter Diagram

The scatter diagram is a way of showing the relationship between two variables, e.g., independent and dependent variables. A scatter diagram graphs individual data points representing an observed pair of values for both X and Y. The scatter diagram may also contain a regression line that fits the data points, along with the equation for the line and the correlation coefficient.

Example 13-6 Scatter Diagram

Serum acetate: propionate ratio is related to serum cholesterol in men but not women (Wolever et al., 1996) (Used with permission)

Independent variables .... Serum acetate: propionate ratio, age, sex
Dependent variables....................Serum cholesterol

As expected, serum cholesterol increased with age for both normolipidemic males ($r = 0.643$, $p < 0.001$) and females ($r = 0.785$, $p < 0.001$), with a relatively even distribution of values across the range of ages (Fig. 13-6).

Figure 13-5: Contribution of food sources to inulin and oligofructose in American diets. Data presented in this figure are from the 1994-1996 Continuing Survey of Food Intakes by individuals using the specialized database for inulin and oligofructose.

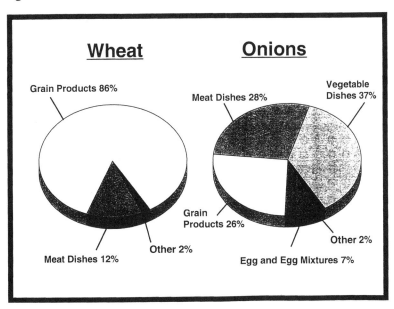

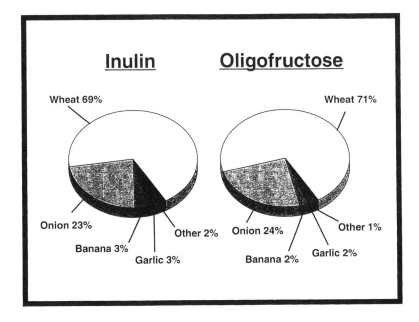

262

Figure 13-6. Relationships between age and fasting serum cholesterol in male subjects (r = 0.643, P < 0.001) and female subjects (r = 0.785, P < 0.001).

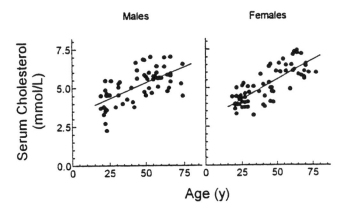

Example 13-7 Scatter plot with regression line and confidence level

The graph demonstrates the relationship between independent (blood glucose) and dependent variables (Blood triglycerides). You can draw a scatter plot with independent variables along the X axis and dependent variables along the Y axis, and then to draw a regression line which shows the slope of the relationship. The equation of the regression line expresses the relationship between x and y (in the form of y = a + bx, where a = the point of interception of the y axis and b = the regression coefficient). You can also draw the line alongside which depict the confidence interval at any point along the regression line.

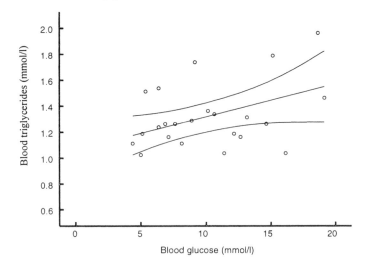

Photographs

A clear photograph can often greatly help readers to understand your message. It must show what you want it to show and as little else as possible. The advice of an informed reader is well worth seeking to ensure that the photographs are of good enough quality, in terms of clarity, focus, contrast, magnification, size, identification of the components under discussion, and, for histopathology, clean white background.

Example 13-8 Photographs (Koh et al. 1994)

Independent variables..... Operation.... sham operation, orchiectomy
Hormone .... sham control, orchiectomy control, estrogen, testosterone
Dependent variables ......Trabecular numbers, bone mass

Fig. 1. Photomicrograph of proximal tibial metaphyseal trabecular bone of sham-orchiectomized rat fed high-fructose magnesium-deficient diet. Modified Goldner stain; 24 x

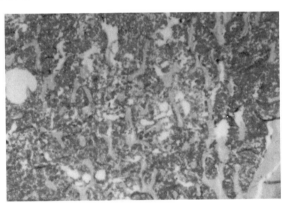

Fig. 2. Photomicrograph of proximal tibial metaphyseal trabecular bone of orchiectomized rat fed high-fructose magnesium-deficient diet. There is no significant trabeculae bone loss as compared to those of sham-orchiectomized rat in Fig. 1. Modified Goldner stain; 24 x

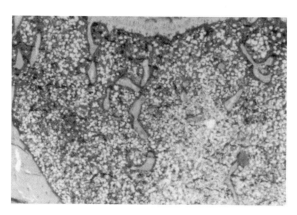

Fig. 3. Photomicrograph of proximal tibial metaphyseal trabecular bone of oestrogen-treated orchiectomized rat fed high-fructose magnesium-deficient diet. There is a significant increase in the numbers of trabeculae as compared to those of sham-orchiectomized rat in Fig. 1. Modified Goldner stain; 24 x

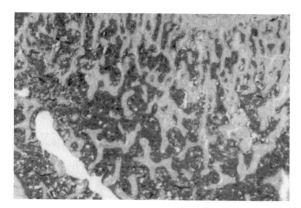

Fig. 4. Photomicrograph of proximal tibial metaphyseal trabecular bone of testosterone-treated orchiectomized rat fed high-fructose magnesium-deficient diet. There is no significant increase of the numbers of trabeculae as compared to those of sham-orchiectomized rat in Fig. 1. Modified Goldner stain; 24 x

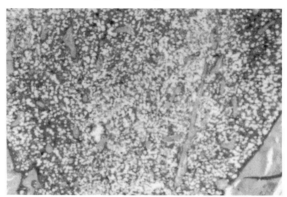

The effects of sex steroids on bone morphology are shown in Figures 1-4. Oestrogen treatment increased the numbers of trabeculae significantly in orchiectomized rats as compared to the sham-control rats (Fig. 1, sham vs. Fig 3, oestrogen). In contrast, testosterone treatment with orchiectomized rats did not increase the number of trabecluae (Fig. 4).

Uniform Requirements for Manuscripts

The 1988 edition of the Uniform Requirements for Manuscripts Submitted to Biomedical Journals should be familiar to authors. It describes the format in which editors agree to receive articles and includes guidelines (Example 13-10) for presenting statistical aspects of scientific research in ways that are clear and helpful to readers.

Example 13-9

Table with Guidelines for Statistical Reporting in Articles for Medical Journals

1. Describe statistical methods with enough detail to enable a knowledgeable reader with access to the original data to verify the reported results.
2. When possible, quantify findings and present them with appropriate indicators of measurement error or uncertainty (such as confidence intervals).
3. Avoid sole reliance on statistical hypothesis testing, such as the use of p values, which fails to convey important quantitative information.
4. Discuss eligibility of experimental subjects.
5. Give details about randomization.
6. Describe the methods for, and success of, any blinding of observation.
7. Report treatment complications.
8. Give numbers of observations.
9. Report losses to observation (e.g., dropouts from a clinical trial).
10. References for study design and statistical methods should be to standard works (textbooks or review papers with pages specified) when possible, rather than to papers in which designs or methods were originally reported.
11. Specify any computer programs used.
12. Put general descriptions of statistical methods in the methods section. When data are summarized in the results section, specify the statistical methods used to analyze them.
13. Restrict tables and figures to those needed to explain the argument of the paper and to assess its support. Use graphs as an alternative to tables with many entries: do not duplicate data in graphs and tables.
14. Avoid non-technical uses of technical terms in statistics, such as random (which implies a randomizing device), normal, significant, correlation, and sample.
15. Define statistical terms, abbreviations, and most symbols.

Further instructions for writing results:

1. The past tense should be used.
2. Subheading can be used for each dependent variable.
3. Statistical symbols should be underlined or italicized.
4. Use the proper case for each statistical symbol.
5. Numerals that start sentences and that are less than 10 usually should be spelled out.

DISCUSSION

The discussion section is an explanation and interpretation of the results. This section is perhaps the most creative part of the experimental type research report. In this section, the researcher discusses the implications and meanings of the findings, poses alternative implications, relates the findings to previous work, and suggests the use of the research results. Most researchers include a statement of the limitations of the study in this section as well, although it can also be found in method or findings. There should be no new findings presented in the discussion section. Any and all things from a statistical standpoint (either descriptive or inferential) should have preceded any material presented in the discussion section. A brief summary of the study may be included in the discussion section of a journal article; a longer, formal summary may be required in a thesis or dissertation.

Key Components of Discussion

The following components are usually included: An example for each component is given using the same paper as Example 13-1, Koh et al. 1994.

1. The major or key finding of your study may be presented at the beginning the discussion section.

A striking finding of the present study is that 15 week orchiectomy of male rats, at which time circulating testosterone was negligible, did not significantly decrease femur weight, lengths, bone areas, numbers of the trabeculae, bone mineral content and bone mineral density. In other words, bone mineral loss did not occur in the orchiectomized rats. Furthermore, testosterone treatment did not increase bone mineral density and the number of trabeculae in the orchiectomized male rats.

2. Discuss your results --- not what you wish they were but what they are.

Unexpectedly, exogenous testosterone treatment of orchiectomized rats decreased body weight significantly compared to the sham control rats (final body weight, 329g vs. 382g).

3. Relate your results back to the introduction, previous literature, and hypotheses. It is usually appropriate to restate your hypotheses in the discussion section and discuss whether they were supported by your data.

If not supported, explain the possible reasons for not supporting. Point out consistencies and inconsistencies of the current results with those in the literature cited earlier in the report, and also explain the reasons for the inconsistencies.

This finding (testosterone treatment of orchiectomized rats decreased body weight) contrasts with our previous observations that testosterone treatment increased body weight significantly in ovariectomized female rats. Oestrogen treatment decreased body weight in both orchiectomized rats in the present study and ovariectomized rats in our previous study. However, body mass decreased to a lesser extent in the orchiectomized rats in the present study than in the ovariectomized rats in our previous study.

4. Explain how your results fit within theory. Important strengths and limitations of your methodology should be discussed.

The most significant finding in the present study is that steroid hormones interacted with nutrients to decrease or increase bone mineral density and the numbers of trabeculae. Testosterone injection interacted with magnesium deficiency and starch to decrease bone mass significantly. The interaction decreased serum calcium significantly. This hypocalcemia may stimulate the secretion of PTH to increase circulating PTH concentration. This results is supported by other studies.

5. Recommend or suggest implications/applications/recommendations of your findings.

The present study indicated that nutrient-sex hormone interactions may be significant in clinical applications of sex hormones for the prevention of osteoporosis in hypogonadal men.

6. Summarize and state your conclusions with appropriate supporting evidence.

An interaction between exogenous oestrogen, dietary fructose, and dietary magnesium deficiency increased serum calcium concentration and reduced circulating PTH levels thereby minimizing bone resorption. Low serum PTH reduced circulating $1,25(OH)_2D$ and urinary cAMP that could also be osteoporotic factors. Exogenous oestrogen interacted with high dietary fructose and dietary magnesium deficiency to increase bone mineral density and the numbers of trabeculae significantly. In contrast to oestrogen, an interaction between exogenous testosterone, starch and magnesium deficiency decreased serum calcium significantly, and thus increased circulating PTH. This increased PTH increased urinary cAMP and circulating $1,25(OH)_2D$ levels. The high circulating PTH, $1,25(OH)_2D$ and urinary cAMP likely caused bone resorption, resulting in decreased bone mass and the numbers of trabeculae.

7. Identify future research area and problems.

Needs further studies to identify mechanisms responsible for interactions of diet and sex hormones to increase or decrease bone mass in ovariectomized or orchiectomized rats. The results should be confirmed in human subjects in the future studies.

Your discussion can further be guided by the following questions taken from the publication manual of the American Dietetic Association:

1. Is the discussion relevant to the findings? Are the results interpreted appropriately and compared with other published data of a similar nature?
2. Are the implications/applications/recommendations logical, well-considered, pertinent yet far sighted?

The conclusion section is a short summary that draws conclusions about the findings of the study in regard to future directions for research or clinical or health care practices.

ABSTRACT

An abstract is a summary that is placed before the introduction in a journal article, thesis, or dissertation. Usually journals and universities put word limits on abstracts, often about 150 words. It varies between 150 to 250 words. The important consideration is to get to the point: What was the problem? Who were the subjects? What did you do? What did you find? What is (are) your suggestion(s) or conclusion(s) based on your findings? (The most useless statement encountered in these abstracts is "Results were discussed.")

Abstracts are also succinct summaries of research that are used as the basis of short paper, oral or poster presentations. In these cases, a brief table may be included, but most authors prefer to use space to present more written information.

270

Basic components of abstract:

1. The research purposes, questions, or hypotheses usually should be made.
2. Highlights of the methodology usually should be mentioned.
3. Highlights of the results usually should be included in the abstract.
4. The abstract for journal article usually should be a single paragraph. (see example 13-10)

However, you should follow the "Guide for Authors" for the particular journal or scientific meeting that you are interested in. For instance, abstracts in the Journal of the American College of Nutrition require four different paragraphs, objective, methods, results, and conclusion. (see example 13-11)

The American Journal of Clinical Nutrition required requires the following five headings:

Background: Provide a sentence or 2 that explains the context of the study.
Objective: State the precise objective, specific hypothesis to be tested, or both.
Design: Describe the study design, including the use of cells, animal models, or human subjects. Identify the control group. Identify specific methods and procedures. Describe interventions, if used.
Results: Report the most important findings, including results of statistical analyses.
Conclusions: Summarize in a sentence or 2 the primary outcomes of the study, including their clinical application, if relevant (avoid generalizations).

Key words should be provided under the abstract for journals, but not for scientific meetings. They are usually limited to five to six words including subjects. But should follow the guideline for each journal.

Example 13-10

 Single Paragraph, Journal of Nutrition

Food deprivation and refeeding influence growth, nutrient retention and functional recovery of rats (Boza et al. 1999) (Used with permission)

The objective of this work was to determine the effects of starvation and refeeding on growth, nutritional recovery and intestinal repair in starved rats. Male Wistar rats, weighing 200g, were starved for 3 d, then refed a soy-based diet for another 3 d. Normally fed rats were given the same diet and used as controls. The variables assessed were as follows: body weight gain and nitrogen retention during recovery after starvation; muscle glutamine concentration; tissue protein content; gut mucosa and liver glutathione levels; intestinal permeability to ovalbumin, lactulose and mannitol; and intestinal tissue apoptosis. Starvation was associated with lower muscle glutamine levels and intestinal muscosa impairment, including a lower content of mucosal protein, a higher level of oxidized glutathione, enhanced permeability to macromolecules and greater numbers of apoptotic cells. Refeeding for 3 d resulted in rapid repair of gut atrophy and normalization of not only intestinal permeability but also of the majority of metabolic markers assessed in other tissues. In conclusion, with the use of severely starved rats, we have established a reversible experimental animal model of malnutrition that might prove useful in comparing the effectiveness of different enteral diets.

Key words: starvation, refeeding, rats, intestinal permeability, glutamine stores

Example 13-11

Multiple Paragraphs, Journal of the American College of Nutrition

Regional differences in consumption of 103 fat products in Belgium: A supermarket-chain sales approach (Hond et al. 1995) (Used with permission)

Objective: In Belgium, important regional differences in mortality exist which have been linked to differences in dietary consumption patterns.
Methods: To study regional food consumption in Belgium, sales data of 103 spreading and cooking fat products in 110 branch stores of a major supermarket chain (Colruyt) for 12 months (1991-92) were analyzed.
Results: Sale of more ordinary and polyunsaturated spreading margarine and of more polyunsaturated low-fat spread in the north in combination with a greater sale of butter and dairy low-fat spread in the south resulted in a P/S-ratio of 0.99 in the north vs. 0.40 in the south ($p < 0.001$) and a U/S-ratio of 1.93 vs. 1.10 ($p < 0.001$) for spreading fats. The P/S- and U/S-ratios of cooking fats were lower in the north (NS). Mortality data correlated positively with the sale of butter and dairy low-fat spread and negatively with sales data of spreading margarine, polyunsaturated spreading margarine, and polyunsaturated low-fat spread (all $p < 0.001$).
Conclusion: Sales data from supermarket chains provide useful information on regional fat consumption and offer interesting perspectives of trends over time.

Key words: nutrition surveys, food habits, dietary fat, epidemiology

Because of word limitation between 150-250 in abstracts, statistical significant levels ($p < 0.01$ or $p < 0.05$) are not usually stated in the abstract.

REPORT OF QUALITATIVE STUDY

Although there is no standard format, the components of a qualitative study are similar to those of other conventional research reports. The first part introduces the problem and provides background and related literature. A description of method is an integral part of the report. The methodology section is usually more extensive than in other forms of research because the methodology is integrally related to the analysis and is also important in terms of validity and reliability. The analysis and discussion section forms the major part of the report. Charts, tables, figures, and photographs are contained in this section and must be integrated into the narrative. The qualitative study strives to provide enough detail to show the reader that the author's conclusions make sense. Some balance between descriptive material of report and conceptual frame work is needed. It has been suggested that 60-70% of the report be descriptive material and 30-40% should be devoted to the conceptual frame work (Merriam, 1988). Qualitative reports are difficult to publish in refereed nutrition or other scientific journals.

The reports of qualitative research generally contain four types of approach:

1. Interpretive
2. Artistic
3. Systematic
4. Theory driven

Summary

The results and discussion sections are written after the data have been collected and analyzed, and the abstract section should be written at the end. The results section is the most important part of the research. This section presents your contribution to knowledge that is what you found. The best way may be to address each of the tested hypotheses: Describe the analysis and results for the first hypothesis first, then describe them for the second hypothesis, and so on. On other occasions, the results may be organized around the independent or dependent variables of interest. The results should be concise and effectively organized and should include appropriate tables, graphs, figures, or photographs.

The discussion section is perhaps the most creative part of the experimental type research report. In this section, the researcher discusses the implications and meanings of the findings, poses alternative implications, relates the findings to previous work, and suggests the use of the research results. A brief summary of the study may be included in the discussion section of a journal article: a longer, formal summary may be required in a thesis or discussion.

Although there is no standard format, the components of a qualitative study are similar to those of other conventional research. The qualitative study strives to provide enough detail to show the reader that the author's conclusions make sense.

274

References

Bailer J.C., Mosteller F. Guidelines for statistical reporting in articles for medical journals. Ann Intern Med 1988; 108: 226

Baumgartner, T.A., Strong, C.H. Conducting and Reading Research in Health and Human Performance. Indianapolis, IN: Brown & Benchmark, 1994.

Beare P.G. Essentials of writing for publication. J Ophthal Nurs Tech 1988; 7: 56-58

Boza J.J., Moennoz D., Vuichoud J., Jarret A.R., Gaudard-de-Weck G., Fritsche R., Donnet A., Schffrin E.J., Perruisseau G., Ballevre O. Food deprivation and refeeding influence growth, nutrient retention and functional recovery of rats. J Nutr 1999; 129:1340-1346

Duffy M. Methodological triangulation: a vehicle for merging quantitative and qualitative research methods. Image J Nurs Scholarship1987; 19: 130-133

Hond E.M.D., Lesaffre E.E., Kesteloot H.E. Regional differences in consumption of 103 fat products in Belgium: A supermarket-chain sales approach. J Amer Coll Nutr 1995; 14:621-627

International Committee of Medical Journal Editors. Uniform requirements for manuscripts submitted to biomedical journals. Br Med J 1998; 296:401

Koh E.T., Tae W-C, Bourdeau J.E., Chung K-W. Oestrogen but not testosterone increases bone density in orchiectomized rats more when fed moderately magnesium-deficient fructose than moderately magnesium-deficient cornstarch. Magnesium Research 1994; 7: 223-232

Koh E.T., Owen W.L., Om A-S. Exogenous oestrogen affects calcium metabolism differently from exogenous testosterone in ovariectomized or orchiectomized rats fed a high fructose diet severely deficient in magnesium. Magnesium Research 1996; 9:23-31

Leedy, P.D. Practical Research: Planning and Design. New York: Macmillan Publishing Co. 1980.

Merriam, S.B. Case Study Research in Education. San Francisco, CA: Jossey-

Bass, 1988.

Monsen, E.R. Research: Successful Approach. Chicago, IL: The American Dietetic Association, 1992.

Morse, J.M. Qualitative Nursing Research. Rockville, MD: Aspen, 1989.

Moshfegh A.J., Friday J.E., Goldman J.P., Chug Ahuja J.K. Presence of inulin and oligofructose in the diets of Americans. J Nutr 1999; 129:1407S –1411S

Okolo, E.N. Health Research Design and Methodology. Boca Raton, FL:CRC Press, 1990.

Patton, M. Qualitative Evaluation and Research Methods. (2nd ed.) Newbury park, CA: Sage, 1990.

Perkin, J. Developing Tables, Graphs, and Figuresmmunicating as Professionals. Chernoff R., ed. Chicago, IL: The American Dietetic Association, 1986.

Pyrczak, F., Bruce, R.R. Writing Empirical Research Reports: A Basic Guide for Students of the Social and Behavioral Sciences. Los Angeles, CA: Pyrczak Publishing, 1992.

Receveur O., Boulay M., Kuhnlein H.V. Decreasing traditional food use affects diet quality for adult Dene/Metis in 16 communities of the Canadian northwest territories. J Nutr 1997; 127:2179-2186

Reynolds, L., Simmonds, D. Presentation of Data in Science. Martinus Nijhoff: The Hague, 1983.

Rossman G.B., Wilson B.L. Numbers and words: combining quantitative and qualitative methods in a single large scale evaluation study. Evaluation Rev. 1985; 9:627-643

Rowles, G.D., Reinharz, S. Qualitative Gerontology: Themes and Challenges. In Qualitative Gerontology. Reinharz, S. Rowles, G.D. eds. New York: Springer-Verlag, 1988.

Sax, G. Principles of Educational Measurement and Evaluation. Belmont, CA: Wadsworth Publishing Co., 1974.

Shayne, C.G. Statistic and Experimental Design for Toxicologist. (3rd ed.) Boca

276

Raton, FL: CRC Press, 1999.

Simmonds, D., Reynolds, L. Data Presentation and Visual Literacy in Medicine and Science. Oxford: Butterworth-Heimann, 1994.

Smith M. Publishing qualitative research. Amer Educ Res 1987; J 24:173-178

Smith R. Steaming up windows and refereeing medical papers. Brithish Med J 1982; 285: 1259-1261

Sugiyama K., He P., Wada S., Saeki S. Teas and other beverages suppress D-galactosamine-induced liver injury in rats. J Nutr 1999; 129:1361-1367

Thomas, J.R. Nelson, J.K. Research Methods in Physical Activity. (3rd ed.) Champaign, IL: Human Kinetics, 1996.

Wilson H.S., Hutchinson S.A. Triangulation of qualitative methods: heidiggerian hermeneutics and grounded theory. Qualitative Health Res. 1991; 1:263-276

Wolever T.M.S., Fernandes J., Rao, A.V. Serum acetate: propionate ratio is related to serum cholesterol in men but not women. J Nutr 1996; 126:2790-2797

Yu S., Morris J.G. Chloride requirement of kittens for growth is less than current recommendations. J Nutr 1999: 129: 1909-1914

Zellmer W.A. How to write a research report for publication. Am.J Hosp Pharm 1981; 38:545-549

CHAPTER 14

===

PUBLICATIONS

Research Report

The previous chapter covered how to organize and write results and discussion sections. Effectively coordinating all this information into a thesis or a dissertation is the major concern of this chapter. In addition, we include information about writing for publication in scientific journals, preparing abstracts, and presenting papers in both oral and poster sessions.

Day stated in his book (1983) that "Scientific research is not completed until the results have been published. Therefore, a scientific paper is an essential part of the research process." Research reports allow the writer to share his/her research findings with other scientists and interested readers. The writer, as communicator, needs to be logical in writing introduction, clear and accurate in describing the methods and relating the results and conclusions of the research.

The major difference between reports of thesis or dissertation and journal articles is the length of the document. Journal articles are limited by established publishing criteria of particular journals in length of 5-10 published pages, whereas there is no page limitation in thesis or dissertation.

THESIS OR DISSERTATION WRITING GUIDELINES

Graduate students completing theses or dissertations are often required by their respective colleges or universities to follow a particular format for organizing and presenting the report.

Thus, students must:

- Collect all documents that outline university, graduate school, and departmental policy for theses or dissertations;

- Review the theses and dissertations of the past graduate students in your department, and be familiar to their formats and styles; and
- Follow the university thesis or dissertation guidelines in writing reports.

Limitations of Chapter Style

Conventional theses and dissertations typically contain four to six chapters. They are mainly Introduction; Literature Review; Materials and Methods; Results; Discussion; and Summary or Conclusions.

The thesis and dissertation also have several introductory pages as prescribed by the institution, usually consisting of the title page, approval page by committee members, acknowledgements, abstract, table of contents, and lists of figures and tables. At the end of the study are the references and one or more appendices that contain items such as subject consent forms, tabular materials not presented in the text, more detailed descriptions of procedures, instructions to the subjects, and raw data.

In other words, the thesis or dissertation is usually arranged in three major parts: preliminary materials, the text or body of the report, and supplementary items.

**Preliminary materials*:*

1. Title page
2. Approval page by committee members
3. Acknowledgments
4. Abstract
5. Table of contents
6. List of tables
7. List of figures

Body of the thesis or dissertation
Chapters:

1. Introduction
2. Review of literature
3. Materials and methods
4. Results
5. Discussion

6. Summary or Conclusions
7. References

Appendices (Supplementary items):

1. Extended literature review
2. Additional methodology, e.g., questionnaires, interview guides, consent form, etc.
3. Additional results, e.g., raw data, summary of ANOVA chart, etc.
4. Other additional materials, e.g., maps of study regions, and other study backgrounds

One-page vita (mainly dissertation)

PRELIMINARIES

Preliminary pages might be necessary for a manuscript, journal article, thesis, or dissertation.

Title Page

The first page of thesis or dissertation is called the title page. The page usually includes the title of thesis or dissertation, relationship of the report to degree requirement, author, name of the institution to which the report is submitted, and date of submission. The purpose of the title is to convey the content of the study, but this should be done as succinctly as possible (see Chapter 3). The title should be typed in capital letters, single spaced, and centered.

280

Example 14.1 Title Page

THE UNIVERSITY OF AMERICA

THE EFFECT OF CALCIUM SUPPLEMENTATION
ON BONE MASS AND BONE FRACTURE

A THESIS
SUBMITTED TO THE GRADUATE FACULTY
In partial fulfillment of the requirements for the degree of
MASTER OF SCIENCE

By
Allen K. Smith
Norman, Oklahoma
1996

Approval Page

The second page is approval page by committee members

Example 14.2 Approval Page

THE EFFECT OF CALCIUM SUPPLEMENTATION
ON BONE MASS AND BONE FRACTURE

APPROVED BY

_____

Allen Johnson, Ph.D.   Chair

_____

Stephen Clark, Ph.D.

_____

Jean Walton, Ph.D.
THESIS COMMITTEE

Copyright

Most theses or dissertations have a copyright. To initiate this, the author must obtain a copyright authorization form (usually available from university) complete it, pay the fee, and include a copyright notice at the front of the thesis or dissertation. The notice appears, centered on a single page, as follows:

Example 14.3 Copyright

COPYRIGHT
By
Stephen Clark
March 1, 2000

Acknowledgements

If author has received a great deal of assistance from a person or persons, an acknowledgement page is included at the beginning of theses or dissertations. Usually, the student's committee members and family members are mentioned.

Example 14.4 Acknowledgement

ACKNOWLEDGEMENT

It is with deep gratitude and appreciation that I acknowledge the professional guidance and friendship of Dr. John Smith. His consistent encouragement and support helped me to achieve my goal. My gratitude goes to the other......

Table of Contents

The table of contents provides an outline of the contents of theses or dissertations. The major headings and subheadings are included along with the page number of each. The table of contents for theses and dissertations will vary according to the style utilized by individual institutions.

Example 14.5 Table of Contents

TABLE OF CONTENTS

LIST OF TABLES... vi

LIST OF FIGURES..vii

CHAPTER
     I.   INTRODUCTION..1-5
     II.   OBJECTIVES...6
     III.   REVIEW OF LITERATURE.....................................7-32
     IV.   METHODS AND MATERIALS..............................33-42
     V.   RESULTS..43-61
     VI.   DISCUSSION..62-73
     VII. SUMMARY...74-75

BIBLIOGRAPHY...76-82

List of Tables and Figures

When you writing your thesis or dissertation you may utilize figures and/or tables. If you do, a separate list for each (tables and figures) should be included after the table of contents. The exact table titles and numbers used in the thesis or dissertation are presented along with the page on which they are located.

Example 14.6 List of Tables

LIST OF TABLES

Table   Page

1. Experimental Groups..34

2. Experimental Water Treatment..36

3. Effects of Fluoride on Bone Mass..42

Example 14.7 List of Figures

LIST OF FIGURES

Figure   Page

1. Photomicrograph of Femoral Trabecular Bone............................46

2. Percent of Calcium in Heart of Mated Mice Comapred
     To Non-Mate Mice...…..…..50

3. Percent of Zince in Tibia of Mated Mice Compared to
     Non-Mated Mice..…..…56

BODY OF THE THESIS OR DISSERTATION

Chapter 1 Introduction

We have discussed introduction in detail in Chapter 3. The purpose of an introduction in a research report is to introduce the problem area, establish its significance, and indicate the author's perspectives on the problem. In other words, introductions usually conclude with an explicit statement of the research hypotheses, purposes, or questions to be answered by the study.

A good introduction requires literary skill because it should flow smoothly yet be reasonably brief. The introductory paragraphs must create interest in the study: thus your writing skill and knowledge of the topic are especially valuable in the introduction. The narrative should introduce the necessary background information quickly and explain the rationale behind the study. A smooth, unified well written introduction should lead to the problem statement with such clarity that the reader could state the study's purpose before specifically reading it (See Chapter 3).

For the master's student, the thesis is the first research effort, and there may be merit in formally addressing such steps as operationally defining terms, delimiting the study, starting the basic assumptions, and justifying the significance of the study.

In a journal article, the introduction is always integrated with the literature review into a single essay. Most institutions of higher education require that introduction and the review of literature in a thesis or dissertation be presented in separate chapters.

Here is an example of introduction from a master's thesis:

Example 14.8 Introduction

The interactive effects of dietary calcium, carbohydrates and sex hormones on parathyroid hormone and bone mass.

Primary osteoporosis is an age-related disorder characterized by decreased bone mass and by increased susceptibility to fractures in the absence of other recognizable causes of bone loss (ref). Osteoporosis is a common condition that affects as many as 15-20 million persons in the United States. About 1.3 million fractures attributable to osteoporosis occur annually in people aged 45 years and older. The cost of osteoporosis in the United States has been estimated at $14 billion annually (ref).

While the risk of developing bone fractures is definitely related to bone mass, thinning of the bones may not by itself be sufficient to result in a fracture. Nutritional and hormonal factors play a key role in the integrity of bone.

Failure to meet the hormonal and nutritional needs of the bone can influence the strength and resistance of bone tissue to fractures.

Estrogen or 17-B-estradiol is the primary circulating hormone in premenopausal women. As women go through menopause the levels of endogenous estrogen drop due to the decline in ovarian function. As these levels of estrogen decrease there appears to be a concomitant acceleration of bone loss. Therefore estrogen is believed to play an anti-resorptive role in bone remodeling.

Testosterone is also believed to contribute to the integrity of bone (ref). Although there are no hormonal events that occur in men comparable to female menopause, studies have reported a fall in plasma testosterone concentration between ages of 50-60 (ref). This fall in plasma testosterone may be a major risk factor for spinal compression fractures in men (ref).

Nutritional factors are also a key in the maintenance of bone. The skeleton is the major reservoir for calcium. Approximately 99% of calcium in the body is stored in the bone (ref). During states of calcium deficiency, calcium homeostasis is maintained at the expense of the skeleton even to the point of producing severe bone disease.

Fructose is a carbohydrate that has been shown to promote calcium phophosphate deposition in the kidneys, increase bone mineral density and bone mineral content in rats compared to other carbohydrate sources. Koh et al. (ref) found that ovariectomized rats, injected with estrogen, and fed a high fructose magnesium-deficient diet had a greater bone mineral density than overiectomized rats, injected with testosterone, and fed a magnesium-deficient starch diet.

Therefore, this study was undertaken to further examine the role of diet and sex hormones in the development or prevention of osteoporosis using rats as an animal model.
The primary questions to be resolved via this study were:

1. To determine whether an interaction of high fructose and sex hormones can prevent bone loss caused from calcium deficiency in female rats.

2. To study a possible mechanism of an interaction of fructose and sex hormones in prevention of bone loss caused by calcium deficiency in female rats.

Chapter 2 Review of Literature

The purposes of literature review are: 1) to identify the problem; 2) to develop hypotheses; 3) to develop the method in your new research; and 4) to build new knowledge. Theoretical formulations from other studies, the major issues of methodology, instrumentation, and interpretation, and background information are presented. A well-organized chapter that shows how the present study may differ from previous ones, and at the same time add to their contributions. Knowledge gaps in the problem area are noted. Through a review of the literature, a theoretical basis and justification for the present study is formed (discussed in the Chapter 2).

Notice that the review is organized around concepts or themes, rather than the one study per paragraph approach. A review of osteoporosis, for example, may be organized into headings on "bone turnover," "effects of estrogen on bone mass," "effects of testosterone on bone mass," "effects of PTH on bone mass," "effects of vitamin D on bone mass," "effects of calcium on bone mass," etc. Several paragraphs under each topic will depict the overall findings and cite individual studies to document the observations. For example, a sentence in one paragraph may be stated as follows: "Several studies have been undertaken which examine the effects of vitamin D on bone in aging women."

The literature review for a journal article should be highly selective, whereas the review for a thesis or dissertation may be less selective. Peripheral research may be cited in a thesis or dissertation when no literature with a direct bearing on the research topic exists. Therefore, this chapter is often the longest one in thesis or dissertation.

Chapter 3 Materials and Methods

Writing method sections has been discussed in Chapter 4. The method section has one major goal. If someone else follows exactly what is described in the methodology section, they should get identical results, and confirm the findings of original researcher. If they do not, then methods were not precise enough or there was a methodological error. Therefore, it is better to over describe than under describe the methods employed, especially in the first draft of a report. The method section contains a description of the physical steps taken to gather the data. Typically, it begins with a description of the subjects, instrumentation and apparatus (measuring tools including models and manufactures). Any additional procedures such as administration of experimental treatments, design and analysis should also be described here. In reporting on completed research, use the past tense. (See Chapter 4)

In the materials and methods section of thesis or dissertation you should describe backgrounds of study, subjects and subject selection procedures, treatment, assays, and instruments used, and statistical analyses in detail using subheading.

1. Backgrounds of the study
2. Subjects' characteristics (e.g. demographic, physical, clinical, etc.)
3. Experimental groups and sample size in each group, usually by table
4. Subject selection criteria (inclusion/exclusion) and procedures
5. Ethics for using human subjects or animals
6. Diet compositions, usually by charts
7. Every step of treatment procedures
8. Data collection procedures
9. Chemical assays
10. Instruments: models, and manufacturer's name and location
11. Statistical analysis, etc.

Chapter 4 Results

We have discussed how to write results section of research report in Chapter 13. This section presents your contribution to knowledge, that is, what you found in your research. The results should be concise and effectively organized and should include appropriate tables, graphs, figures, and/or photographs. Here are some suggestions:

- Don't repeat what is already clear to the reader from reviewing the tables and figures, which, if well-constructed, will show both the results and the experimental design.
- Present the data or results obtained from the study in a straightforward, factual manner, without commentary or interpretation.
- The Results section must account for all subjects who actually entered the study and not only those who stayed in it as assigned.
- Write this section in the past tense.

Chapter 5 Discussion

As results, we have already discussed about discussion section in Chapter 13. The discussion section is perhaps the most creative part of the experimental type research report. In this section, the researcher discusses the implications, limitation, and meanings of the findings, poses alternative implications, relates the findings to previous work, suggests the use of the research results, and identifies the future research areas. Here are some suggestions:

- Write this section in the **present tense**; the findings of the paper are now considered established as scientific knowledge.
- Discuss controversial issues clearly and fairly.
- Stress, rather than conceal, anomalous results for which no explanation is readily available. You can be sure that if you don't mention them, a reviewer will.
- It's all right to speculate; just make sure that any theories you include are appropriate.
- Avoid unqualified statements and conclusions not supported completely by the data.
- Include suggestions for future study when appropriate.
- Review how your study fits into or changes the established body of knowledge.

Chapter 6 Summary

This is the shortest chapter in a thesis or dissertation. A brief summary of the study including brief methodology and major findings and general conclusions and recommendations resulting from your research can be placed in this chapter. The writer must be sure not to add anything new here because it is an account of what has already been written in the report.

Each chapter is to start on a new page.

References

The body of the text is always followed by an alphabetized list of references. It should be presented in accordance with whatever style format is required by the student's university.

Appendices

The appendices serve as the repository for supplementary materials that are unnecessary for inclusion in the body of the text, but can be used by the reader to clarify various aspects of thesis or dissertation. For instances, questionnaires, interview guides, consent form, raw data, etc., can be included.

PRESENTATIONS AND MANUSCRIPTS

Original research, in addition to your thesis or dissertation, can be presented in many ways: technical reports, abstracts, posters at scientific meetings, oral presentations, video and film presentations, and manuscripts prepared for publication in professional and research journals. Graduate students usually have some obligations for presenting their research outcomes at the scientific meetings. You can make either oral presentation or poster presentation. To present your paper at the scientific meeting, your abstract should be accepted by the organization.

Oral Presentations

First, you have to answer for the following question, "What do I want my audience to know when they leave the room?" Your answer may be "I want to inform…" Before your presentation write a statement of purpose. It will help you remain focused.

Oral presentations are frequently orchestrated with supportive slides, overheads, and power point. These visual materials are designed to the listener. They should be readable from anywhere in the room; thus, they should be simple and composed of large letters and numbers, simple graphs and/or figures. A useful guide for preparing effective slides recommended by the Experimental Biology (former Federation of American Societies for Experimental Biology) is to present no more than 42 characters and spaces horizontally and no more than 14 lines (including double-spaced lines) vertically. If you use different colors for graphs, you should not use more than four colors in a slide. You may use a blue background, yellow 22-24 point headlines, and white 16-18 point text.

Too much data overwhelms listeners; therefore, significant results should be focus of an oral presentations. Again, keep visual aids as simple as possible, using one point per visual. Using seven to ten slides for 15 minutes' presentation would be recommendable.

The best way to keep the audience's attention is to tell them what's in it for them at the beginning. Therefore, a brief review of the literature, providing a rationale for the research being presented, is the most important part of presentation. If they understand the study rationale and importance at the beginning, they will listen more carefully to the information being offered. They will want to know about subject selection criteria, data collection methods, data analysis methods, results of the study, and what conclusions can be drawn from the information presented to them.

To make a best presentation, arrive at the location of your presentation at least one hour before you are to speak. Check to make sure all the equipment you will need is there and is in working order.

When you look up from your text, don't just look straight ahead; make eye contact. Practicing your talk will allow you to look up at the audience more frequently. Making eye contact helps relieve speaker anxiety and helps the audience stay focused on your message.

Audience members can comment on similar research or findings, pose questions about current research, and offer suggestions for future research. They can also challenge with you. When they ask you questions, don't answer the question too quickly. Be sure to give the questioner a chance to ask the entire question before jumping in with an answer. Do not answer too much. Be brief and to the point, emphasizing again your main point.

Here are some additional tips for oral presentations:

- Practice giving your presentation several times before the seminar date. **Practice, practice, practice.**
- Create a positive focus
- Affirm and visualize your success
- Remember that you are a professional presenting scientific information to a professional audience.
- Dress in a neat, attractive professional manner, as you wish to instill respect in your audience.
- Make sure that slides are in proper sequence according to your presentation. Once your presentation begins, you want nothing to occur which will divert the audience's attention away from the topic at hand.
- Try to appear poised. Stand calmly; do not fidget with rings, play with your hair, rub your chin, shuffle your feet, etc. Again, you want the audience to feel you are in complete control. On the other hand, try act natural and not stand totally immobile or as if you were made of stone.
- Do not try to arouse the sympathy of the audience by commending informally about being nervous, etc. You want to give the impression that you know exactly what you are doing and that you are both confident and capable.
- Capture the full attention of your audience in **the very beginning**.
- Stay attuned to your audience. Watch their faces for reactions.
- Know your subject well enough so that you can talk about it in a conversational tone.
- When you become nervous (either before or during) the presentation, try to relax and remember that you are the expert in the crowd.
- Be sure and provide a **meaningful summary and conclusion**.

Poster Presentations

Many professional and scientific organizations support both short oral presentations (about 15 min) and poster sessions. A poster at a scientific meeting is an enlarged graphic display containing a title, the authors' names, and text and figures explaining research. The presentation of poster is different from that of short oral presentations. Posters are more visual than oral presentations and allow for more interaction between the author and the audience. The difficult part is to arrange the material so that people standing behind others can read readily. The abstract, brief and clear study objectives, brief materials and methods, key data, summary or conclusions and recommendations are particularly important for presentation. Posters can be illustrated with graphs, tables, figures, photographs, and illustrations.

Suggestions from the experts (James)

- Typefaces and sizes. Do not use all upper case type. Use upper and lower case. Titles should be bold between 30 and 36 points. Make the title clear and concise. Viewers should be able to read your title from 15-20 feet away.
- Author's affiliation should be 30 points. First level headings should be 30 points. Second level headings should be 24 points. Body copy should be 20 points.
- Graphics should dominate. Include about 50% white spaces. For clarity remove all non-essential information. Use active voice; it is shorter and more direct. Be consistent with terms used in text and figures.
- Limit your information.

Manuscripts

Selecting the journal that might publish the paper should be one of the first steps in getting ready to write. The author reads the literature and becomes knowledgeable about journals that would likely to publish a particular topic. Tailoring the format of the manuscript to suit the chosen journal and following the set guidelines for authors are important for submission to that journal. For instance, if you have a basic animal study, it would be better fitted in the Journal of Nutrition, and if you have a clinical study, it would be fitted in the American Journal of Clinical Nutrition.

The body of thesis or dissertation should be a complete research report using the appropriate style for the journal to which it is to be submitted. Included are the introduction, review of literature, materials and methods, results, discussion, references, tables, and figures. You should shorten the paper by focusing on the significant findings to fit in the bounds set by the journal. This is typically 15 to 25 double-spaced pages for single experiments, whereas thesis or dissertation would be a few times longer. You must greatly reduce the first two chapters, introduction and literature review, to fit in a few paragraphs of introduction in journal articles.

Formats for published articles typically require the following items:

1. Title page
2. Author(s)
3. Organizational affiliation
4. Introduction
5. Materials and Methods
6. Results
7. Discussion
8. Abstract
9. References

Author's Guidelines

If you write a manuscript for a particular journal, you should follow the authors' guideline or the guidelines for manuscript review for that journal. By reading authors' guidelines you could have better idea for writing research reports. You will learn what they emphasize in their manuscripts from the authors' guidelines.

We will go over the authors' guidelines of **the American Journal of Clinical Nutrition,** and the **Journal of Nutrition.** You will learn what they emphasize in their manuscripts.

Submitting the article

Many journals provide a check list for the author to ascertain that every-thing required for submission is included with the manuscript. All required paper work should be reviewed for completeness.

Each journal requires an original and several copies of the manuscript. In peer-reviewed journals, two or more reviewers read and critique each paper and they each need a copy of the paper. The experts read the manuscript for scientific accuracy, clarity, completeness, and contribution of new knowledge. Their criticism and comments are designed to make the paper clearer and more complete.

Eventually, you will receive a letter from the editor. It may indicate full acceptance, acceptance with minor revisions, the need for extensive revisions with another peer review, or rejection.

Acceptance rates vary from 10 percent for very prestigious and specialized research journals as high as 70 percent for less prestigious and practical journal. Do not be discouraged if your paper is rejected. Every body has had papers rejected. Carefully evaluate the reviews and determine whether the paper can be published in another journal. Do not send the paper out another journal without evaluating the reviews and making appropriate revisions.

Following are the three most common reasons that journals reject articles.

1. The manuscript is inappropriate for the particular journal and its specific audience. Obviously many writers do not read the journal's mission statement and the type of articles the journal accepts in authors' guidelines of the journal.
2. The manuscript describes poorly designed or poorly conducted studies. Following are examples: insufficient information, inadequate samples, biased samples, confounding factors, vague endpoints, straying from the hypothesis, poor control of numbers, etc.
3. The manuscript is poorly written. Anytime reviewers notice grammatical, typographical and/or stylistic lapses in the manuscript, their attention is drawn away from your message, and comprehension suffers.

There usually is a lag of several months between final acceptance of the article and publication. During this period, the manuscript is copy edited for style, grammar, and conventions specific to the journal.

Summary

Research reports allow the writer to share his/her research findings with other scientists and interested readers. The writer, as communicator, needs to be logical in writing introduction, clear and accurate in describing the methods and relating the results and conclusions of the research. A typical format for thesis or dissertation has been presented, but this may be modified according to the particular requirements of the students' university. Information regarding the difference between the format of thesis or dissertation and journal article has been discussed.

Original research, in addition to thesis or dissertation, can be presented in many ways: technical reports, abstracts, poster or oral presentation, video and film presentations, and manuscripts prepared for publication in professional and research journals.

If graduate students write a manuscript for a particular journal, they should follow the authors' guideline of that particular journal. Through the guidelines students must learn how to prepare their manuscripts for the most prestigious nutrition journals.

References

Ary, D., Jacobs, L., Razavieh, A. Introduction to Research in Education. Holt, New York: Rinehart & Winston, 1985.

Baumgartner, T.A. Strong, C.H. Conducting and Reading Research in Health and Human Performance. Dubuque, IA: Brown & Benchmark, 1994.

Beare, P.G. Essentials of Writing for Publication. J Ophthal Nurs Tech. 1988.

Booth, V. Communicating in Science: Writing a Scientific Paper and Speaking at Scientific Meetings. (2nd ed.), Cambridge, England: Cambridge University Press, 1993.

Boza J.J., Moennoz D., Vuichoud J., Jarret A.R. Food deprivation and refeeding influence growth, nutrient retention and functional recovery of rats. J Nutr 1999; 129:1340-1346

Briscoe, M.H. A Researcher's Guide to Scientific and Medical Illustrations. New York: Springer-Verlag, 1990.

Campbell, W.G. Form and Style in Thesis Writing. Boston, MA: Houghton, Mifflin Co, 1970.

Campbell, W.G., Ballou, S.V. Form and Style: Theses, Reports, Term Papers. Boston, MA: Houghton Mifflin, 1978.

Chernoff, R. Writing Journal Articles. In Communicating as Professionals. Chernoff, R. ed. Chicago, IL: The American Dietetic Association, 1986.

Day, R.D. How to Write and Publish a Scientific Paper. (2nd ed.) Philadelphia, PA: ISI Press, 1983.

Day, R.D. How to Write and Publish a Scientific Paper (4th ed.) Phoenix, AR: Oryx Press, 1994.

Guide for Authors, The Journal of Nutrition. J Nutr 2000; 130: 1-3

Howard, G. Basic Research Methods in the Social Sciences. Glenview, IL: Scott, Foresman and Co., 1985.

Huck, S.W., Cormier, W.H., Bounds, W.G. Reading Statistics and Research.

New York: Harpers & Row Publishers, 1973.

Huth, E.J. Medical Style and Format: An International Manual for Authors. Philadelphia, PA: ISI Press, 1987.

James, D.S. Writing and Speaking for Excellence: A Brief Guide for the Medical Professional. Sudbury, MA: Jones and Bartlett Publishers, 1998.

Kidder, L. Research Methods in Social Relations. New York: Holt, Rinehart & Winston, 1981.

Knight K.L., Ingersoll C.D. Structure of a scholarly manuscript: 66 tips for what goes where. J Athletic Training 1996; 31: 201-206

Kris-Etherton, P.M. Developing Oral Presentations. In Communicating as Professionals. Chernoff, R. ed. Chicago, IL: The American Dietetic Association, 1986

Leedy, P.D. Practical Research: Planning and Design. New York: Macmillan Publishing Co., 1974.

Mitchell, J.P. The New Writer: Techniques for Writing Well with a Computer. Redmond, Washington: Microsoft Press, 1987

Monsen, E.R. Research: Successful Approaches. Chicago, IL:The American Dietetic Association, 1992.

Morra M.E. How to plan and carry out your posta session. Oncol Nurs Forum 1984; 11:52-55

Pyrczak, F., Bruce, R.R. Writing Empirical Research Reports: A Basic Guide for Students of the Social and Behavioral Sciences. Los Angeles, CA: Psyczak, 1992.

Publication Manual of the American Psychological Association. (3rd ed.) Washington, D.C.: American Psychological Association, 1983.

Relman, A.S. "Journals". In Coping with the Biomedical Literature. Warren, K.S. ed. New York: Praeger Publications, 1981.

Schmid, C.F. Schmid, S.E. Handbook of Graphic Presentation. (2nd ed.) New York: John S. Wiley, 1979

Simmonds, D. Charts and Graphs: Guidelines for the Visual Presentation of Statistical Data in the Life Sciences. Baltimore, MD: MTP Press, 1981.

Strunk, W. Jr., White, E.B. The Elements of Style. (3rd ed.) New York: Macmillan Co, 1979.

Teel C.S. Completing the research process: presentations and publications. J Neurosci Nurs 1990:22: 125-127

Thomas, J.R., Nelson, J.K. Research Methods in Physical Activity. Champaign, IL: Human Kinetics, 1996.

Tornquist E.M. Strategies for publishing research. Nursing Outlook, 1983:31:180-183

Trelease, S.F. The Scientific Paper: How to Prepare It, How to Write It. (2nd ed.) Baltimore, MD: Williams and Wilkins Co, 1951.

Tufte, E.R. The Visual Display of Quantitative Information. New York: Graphics, 1983.

Zellmer, W.A. How to write a research report for publication. Am J Hosp Pharm 1981:38:545-549

CHAPTER 15

==

WRITING THE RESEARCH PROPOSAL

Introduction

One of the goals of this book is to help prepare a graduate student to develop a research proposal. The research proposal components are an abstract, specific aims, introduction (background and significance; and supporting evidence), and methods. Chapters 1 through 4 in this text pertain specifically to the body of the research proposal. Other chapters relate to various facets of planning a study: hypotheses, designs, measurements, and statistical analyses. This chapter attempts to bring the proposal together.

To develop a research proposal a student should progress through the following steps: identifying the research problem; choosing the topic; generating the hypotheses; and developing the research protocol.

Identifying the Research Problem

Of the many major issues facing a graduate student, a primary one is the identification of a research problem. Problems may arise from real-world settings or be generated from theoretical frameworks. The source of research problems will vary according to the experience of the person contemplating an investigation, but it is generally agreed that the process begins with a question or need. Curiosity is a good motivational factor in identifying the research problem.

Before beginning to write a research proposal, some general as well as specific reading should be done. From this reading a general knowledge of the topic can be gained. This knowledge should include a familiarization with areas of controversy, designs, methods, and characteristics of the subject studied, as well as those not yet studied, and recommendations made by others.

Choosing the Topic

The most important element of a research study is defining the research question or problem. The first part of the research process is formation of the idea, topic, or research question. A researcher can select a topic in a variety of ways including coming across evidence of a problem that needs to be solved, encountering unanswered questions in the literature, or personally experiencing situations leading to unanswered questions. The purpose of the topic is to convey the study, but this should be done as succinctly as possible. In another word, the cause- and effect-relationship of the independent and dependent variables should be described in the topic.

Once the topic has been selected, the study question should be defined as precisely as possible. The question should not be too vague or too broad. To move from the selection of a broad topic area to framing a concrete research problem, the following questions can help guide your thinking:

- What about this topic is of interest to me?
- What about this topic is of relevant to my practice?
- What about this topic is controversial and unsolved in the literature?
- What about this topic is of importance or significance in social or scientific aspects?

By answering these reflective questions, you will begin to narrow focus on a researchable problem.

Formulating the Hypotheses

After you have stated the research problem, you must present the hypothesis. The hypothesis is the expected result. When a person sets out to conduct a study, he or she generally has an idea as to what the outcome will be. This anticipated solution to the problem may be based on some theoretical construct (Deductive reasoning and Inductive reasoning), on the results of previous studies, or perhaps on the experimenter's past experience and observations. The research should have some experimental hypothesis about each sub-problem in the study.

Developing the Research Protocol

Protocol development follows hypothesis generation. Preparation of a written protocol is essential. It gives the researchers a chance to think through all of the stages necessary to carry out the project. It is necessary to write down the essential information in each section of the protocol to ensure that all aspects of the study are covered.

Funding organizations frequently specify in program guidelines what information should be included in a project description and how it should be presented. The National Institutes of Health (NIH), for example, organizes the narrative into specific aims, significance, progress report/preliminary studies, and method sections. Other organizations may request different formats (which should be followed carefully), but the information necessary for a proposal narrative is much the same for all proposals.

Many colleges and universities offer internal funding for graduate student research. Although these grants usually are not large, they do offer funds for research. These typically require a two- to five-page proposal that has the support of the major professor and often the department chair. The contents usually include an abstract, a budget, and short narrative focusing on the proposed methodology and why research is important.

The number of pages of a protocol varies. For the National Institutes of Health (NIH), the maximum number of page is 25, and for the USDA Cooperative State Research Competitive Research Grants, the limit is 20. Although the suggested page length may vary, the essential components of the protocol do not vary. These components are discussed below.

Grant proposals include:

- Abstract
- Specific Aims
- Background and Significance
- Preliminary Studies and Progress Report
- Research Design and Methods
- Time Frame
- Budget
- Vitae of Investigators

Abstract

The abstract of the protocol should be written exactly as it will be written for a scientific meeting without the results and discussion section. The abstract should condense the narrative sections of the protocol, briefly describing the project and its significance, and summarizing methodology you plan to use. The abstract creates the first impression most reviewers have of your proposed work, and the only impression for some reviewers. Some agencies use information contained in the abstract to assign the proposal to review groups. The abstract should be written in lay terms, since it may be used for publicity purposes by the funding agency.

The format of NIH describes this section as follows, "State the application's broad, long-term objectives and specific aims, making references to the health relatedness of the project. Describe concisely the research design and methods for achieving these goals. Avoid summaries of past accomplishments and the use of the first person. This description is meant to serve as a succinct and accurate description of the proposed work when separated from the application. If the application is funded, this description, as is, will become public information."

Specific Aims

The specific aims can be a substitute for the research question in protocol development. The specific aims section should never exceed one page. This section usually contains one or two short paragraphs about the background leading up to the development of the study. Next the specific aims are listed numerically.

You should provide answers to this basic question about your project in this section:

- What do you plan to do?
  Explain what you plan to accomplish with your research. Describe your project clearly; set realistic goals.

The format of NIH grant application (PHS 398) contains the following statement in the specific aims section, "List the broad, long-term objectives and what the specific research proposed in the application is intended to accomplish. State the hypotheses to be tested." **The Specific Aims section of the research** plan is the most critical page of the entire proposal.

Introduction

The introduction consists of two parts: (1) **Background**, which is a short review of the pertinent literature that has been published on the subject; and (2) **Supporting evidence**, which is a summary of relevant work published by the investigator or his or her coworkers.

The introduction part of NIH grant application contains two sections: (1) Background and Significance; (2) Preliminary studies and Progress report. We will discuss the introduction section with the NIH format.

Background and Significance

Briefly sketch the background leading to the present application, critically evaluate existing knowledge, and specifically identify the gaps which the project is intended to fill. State concisely the importance and health relevance of the research described in the application by relating the specific aims to the broad, long-term objectives. Two to three pages are recommended.

You should answer this basic question about your project:

- Why is this work important?
  Place your particular project in perspective and document the need for this work by demonstrating an awareness of other related research.

A well-designed diagram at the beginning of the Background and Significance section may reveal the general theory and hypothesis and their tests at a glance, and will be much appreciated by reviewers.

Preliminary Studies/Progress Report

Should provide an account of the principal investigator's preliminary studies pertinent to the application information that will help to establish the experience and competence of the investigator to pursue the proposed project. If the novice has not published at all, he or she should simply name this section

"Introduction," without dividing it into two sections. Regardless of whether the researcher has published previously, there needs to be a current review of the literature. At the very least, it is necessary to search back 5 years in the literature on the subject. If possible, you should include articles from peer-reviewed journal only.

You should provide answers to the following question about your project:
- What have you already done on the project?
  Include results of any preliminary studies and related work that you have done in this area, and indicate how the proposed work builds on the earlier research.

Research Design and Methods

The methods section has one major goal. If someone else follows exactly what is described in this section, they should get the identical results as the original researcher. If they do not, then the methods were not precise enough or there was a methodological error. You should provide the following specific information in detail.

Describe the research design and the procedures to be used to accomplish the specific aims of the project. Include how the data will be collected, analyzed, and interpreted. Describe any new methodology and its advantage over existing methodologies. Discuss the potential difficulties and limitations of the proposed procedures and alternative approaches to achieve the aims. As part of this section, provide a tentative sequence or timetable for the project. Point out any procedures, situations, or materials that may be hazardous to personnel and the precaution to be exercised.

You should answer this basic question in this section:
- How do you plan to accomplish your objectives?
  Describe the methodology you will use to conduct your research. Be specific in outlining procedures. Include a schedule showing expected progress on your project. Address potential difficulties in data collection and/or analysis and describe how you propose to deal with them.

The methods section usually includes the following subheadings:

- Subjects : patients and animals
- Inclusion and exclusion criteria

- Recruitment strategy
- Diet or other interventions
- Procedures/measures
- Statistics
- Time table
- Ethics: Human subjects/animals
- References

Why Grant Proposal Fail

An analysis of why NIH grant applications for clinical research fail was made by Janet Cuca of the Division of Research Grants at the NIH and reported in Cancer Investigation. She found the reasons most grants are disapproved or receive low priority scores are problems with the following:

Experimental Design. 65%: Technical methodology is questionable, unsuited, or defective. **41%:** Data collection procedures are confusing or use inappropriate instrumentation, timing, or conditions. **40%:** Study group or controls that are of inappropriate composition, number, or characteristics. **31%:** Data management and analysis that are vague, unsophisticated, and not likely to provide accurate and clear-cut results.

Research Problem. 47%: Hypothesis is ill-defined, lacking, faulty, diffuse, or unwarrented. **30%:** Proposal is unimportant, unimaginative, or unlikely to provide new information.

Investigator. 17%: Principal investigator has inadequate expertise or familiarity with literature in the research area, poor past performance, poor productivity on an NIH grant, or insufficient time to be devoted to the project.

Resources. 4%: Inadequate institutional setting, support staff, laboratory facilities, equipment, or personnel; restricted access to appropriate patient population; insufficient collaborative involvement of colleagues and co-investigators.

Suggestions for Grant Writing

- Design a concise, yet inclusive title. Request a copy of the titles of recently approved applications. Look closely at format and style.

- Stay within the maximum page allowance. Reviewers may not review your application if it exceeds the maximum page allowance.
- Enclose references related to the research subject or methods. You should convince the reviewer you know the most relevant and pertinent information on your research question.
- Be realistic and explicit when it comes to the budget. Itemize all costs exactly, including animals, housing, supplies, equipment, and personal.
- Discuss in the Methods of Procedure section your approach to answering each question or specific aim.
- Consult a biostatisticians if you don't fully understand statistics.

How to Prepare a Strong Proposal

The formal research proposal should be carefully prepared. The pragmatic definition of a strong proposal is simply any proposal that gets funded. Explanations for success or the lack of it are legion. Unfortunately, good science does not always lead to a successful proposal, although badly written proposals are often funded if their science and the principal investigator's background are sufficiently strong. On the other hand, proposals based on faulty science are hardly ever successful. Anyhow, without exception, the proposals could be stronger (get a better priority score) if they were better written. By "better written" we mean easier for the reviewer to read and to understand. The first impression that a reviewer gets of a proposal should be impressive. The following are some recommendations for the strong proposal:
- Proper use of a word processor should eliminate typographical errors from the text.
- Every page of a proposal should have the same general appearance.
- Good writing is brief: Reduce the text sufficiently that the page limitations can be met.
- The budget must be reasonable at a glance.
- The Biosketch (vitae of investigators) should indicate solid training, steady productivity, and recent publications pertinent to the proposed research.
- Make a good impression with figures in the Progress Report – Preliminary Data section.
- Exposition is clear, logical, and brief.
- Scientific content should be strong.

The appearance of the proposal, how it is assembled, its neatness, and how closely it resembles published material all have an impact on the way an investigator is perceived by the reviewers.

Summary

One major goal of this chapter is to help prepare graduate students to develop a research proposal. The research proposal is essentially the plan for the study. The proposal components are an abstract, specific aims, introduction (background and significance; and supporting evidence), and methods. To develop a research proposal a student should progress through the following steps: identifying the research problem; choosing the topic; generating the hypotheses; and developing the research protocol.

Funding organizations frequently specify in program guidelines what information should be included in a project description and how it should be presented. Although the grant application formats of various funding agencies may vary, the information necessary for a proposal narrative is much the same for all proposals. In your proposal narrative you should provide answers to the following basic questions about your project: What do you plan to do? Why is this work important? What have you already done on the project? How do you plan to accomplish your objectives?

The pragmatic definition of a strong proposal and how to write a strong proposal have been discussed. The appearance of the proposal, how it is assembled, its neatness, and how closely it resembles published material all have an impact on reviewing the project.

308

References

Application for a Public Health Service Grant (PHS 398) Includes Research Career Awards and Institutional Research Service Awards. Rockville, MD: National Institutes of Health, 1998

Cuca, J. Why clinical research grant applications fare poorly in review and how to recover, Cancer Investigation 1987; 5: 55-58

Depoy, E., Gitlin, L.N. Introduction to Research. St Louis, MO: Mosby, 1993.

Fredrickson, D.S. Biomedical research in the 1980s. N Engl J Med 1981; 304:509-17

Gardner, D.C., Beatty, G.J. Dissertation Proposal Guidebook: How to Prepare a research Proposal and Get it Accepted. Springfield, IL: Charles C. Thomas Publisher, 1980.

Ireton-Jones, C.S., Gottschlich, M.M., Bell, S.J. Practice-Oriented Nutrition Research: An Outcomes Measurement Approach. Gaithursburg, MD: Aspen Publishers, Inc., 1998.

Locke, L.F., Spirduso, W.W., Siverman, S.J. Proposal That Work: A Guide for Planning Dissertations and Grant Proposala (3rd ed.) Newbury Park, CA: Sage, 1993.

Madsen, D. Successful Dissertations and Theses: A Guide to Graduate Student Research from Proposal to Completion (2nd ed.) San Francisco, CA: Jossey-Bass Publishers, 1992.

Marshall, C., Rossman, G.B. Designing Qualitative Research. Newbury Park, CA: Sage 1989.

Ogden, T.E., Goldberg, I.A. Research Proposals: A Guide to Success. (2nd ed.) New York: Raven Press, 1995.

Thomas, J.R., Nelson, J.K. Research Methods in Physical Activity. (3rd ed.) Champaign' IL: Human Kinetics, 1996.

PART V

==

USING COMPUTERS IN RESEARCH

Part V highlights the use of computers in research and demonstrates various examples of research data analysis.

Chapter 16 discusses hardware, and software. We explain data entry, graphical, and statistical software. Data analysis examples are: 1) summary statistics; 2) a simple histogram; 3) a t test for independent groups; 4) chi-square analysis of a contingency table; 5) calculation of correlation coefficients; 6) simple linear regression; 7) multiple regression; 8) one way analysis of variance without repeated measures; 9) two factor analysis of variance for independent groups; 10) single factor analysis of variance with repeated measures; and 11) two factor analysis of variance with repeated measures.

CHAPTER 16

===

USING COMPUTERS

Introduction

As recently as twenty years ago many research projects did not utilize computers. Today virtually all research projects involve the use of computers. Computers are used to collect data, store data, analyze data, and write reports. There are two types of computers in use in research: specialized computers and general-purpose computers. Specialized computers are used as part of many data collection systems. These specialized computers are similar to the computers used in new automobiles to monitor the operation of the various parts of the car. Although these computers are extremely common they will not be discussed here because their use depends completely on the equipment to which they are attached. General-purpose computers can be programmed to do a variety of jobs. General-purpose computers include the personal computers that have become so familiar. In this chapter we will discuss the parts, operation, and use of general-purpose computers in a research project. From here on, since only general-purpose computers will be discussed, the word computer will be used to mean general-purpose computer.

Hardware

In order to be a computer a device must have four parts: an input component, an output component, a central processing unit (CPU) that includes memory, and a storage component. Currently, the input component is most frequently a keyboard and the output component is a monitor. Originally, monitors were called cathode ray tubes (CRT). While there are still occasional references to CRTs this terminology has almost disappeared.

The CPU is the heart of the computer. It manages the data, performs the operations, and controls the other components of the computer system. In PC computers the CPU is usually a "computer chip" which is mounted on a "mother board" inside the computer. CPUs differ in their ability to access data and in the speed with which they access data. Just a few years ago CPUs could access 64,000 individual pieces of data. Today CPUs can address several million individual pieces of data. The amount of data that can be accessed will certainly continue to increase. The speed of computer chips is measured in several ways. The most common measure of speed for chips in personal computers is megahertz (MHz). Currently chips in personal computers operate at up to 500 MHz. The speed of chips will also surely increase. The speed of the chip is the number of "computer operations" which the chip can do in one second. Because it requires several "computer operations" to complete one arithmetic operation the perceived speed of a computer depends not only on the chip but also on the other components of the computer as well as the software being used.

The memory of a computer, sometimes called random access memory (RAM), is the portion of the computer in which the current instructions for the computer are stored and in which the operations are carried out. One way to think about memory is as follows. Suppose that you are given a set of instructions and are asked to carry out a complex calculation using the instructions. If the instructions are written on 3 x 5-inch index cards and if you must do the calculations on other 3 x 5-inch index cards then the calculations will take you a long time. You will need to find the card containing the next portion of the instructions then read the instructions. You will then need to continue your calculations by writing on the cards, probably looking back at previous cards from time to time. On the other hand, if the instructions are on a large sheet of paper and if you are writing on a large sheet of paper the calculations will obviously go much faster. The memory inside a computer works like the cards and paper in this analogy. Clearly it is desirable to have as much memory as possible.

Both data and programs are stored in files that can be read by the CPU. Several different types of storage devices are used to hold these files. Early computers used IBM cards to store programs and data. At this time programs and data are stored in files stored magnetically either on disks permanently attached to the computer, called hard disks, or on floppy diskettes of various sizes. These disks are highly reliable and dependable; however, it is still necessary to keep multiple copies of important files. The fact that both programs and data are stored so that the CPU can access them quickly is one of the things that make computers work as well as they do. Program files and data files have different characteristics so that the CPU can distinguish between them. When instructed to do so the CPU loads a program into memory and begins executing the individual program instructions one by one. The program accesses the data files that are stored on the disks, as they are needed. In fact, different programs store data in different ways.

Software

There are two fundamental types of software: operating systems (OS), and applications. Each of these fundamental types is divided into several sub classifications. Every computer must have an operating system. The operating system is the software that controls all the components of the computer. When you direct the computer to print the direction goes through the operating system that sends the instruction to the printer. Similarly when you direct the computer to save something as a file the direction goes through the operating system that sends the instruction to the disk on which the file is being saved. The operating system is the software that gives the computer its "personality". Currently the most common operating system for personal computers is Microsoft's Windows. Other operating systems in common use are Microsoft's DOS and OS/2, UNIX, and its derivative LINIX.

Applications software is what makes the computer do what you want it to do. Examples of applications software include word processing programs, spreadsheet programs, database programs, statistical programs, and, of course, games. Applications software is written to work with a particular operating system on a particular type of computer. Without question applications software is the most important part of any computer system. Some people who have used computers for years still believe that the applications software that they use is the computer. Whenever a computer purchase is contemplated the first consideration should be the applications software that will be run on the machine. Many people have wasted a considerable amount of money and experienced substantial frustration because they bought a computer and operating system that would not run the applications that they wanted to run. Most research projects will require the use of four different types of application software: data entry software, graphical software, statistical software, and word processing software. These types of software need to work together in order to analyze the data and write a report. Sometimes a single program will be able to carry out all the required functions; however, usually more than one program will be required.

Data Entry Software

There are programs that are intended only for data entry; however, most research projects that use personal computers use either a spreadsheet program or a database program for data entry. Spreadsheet programs divide the computer screen into rows and columns. The intersection of a row and a column forms a cell. Data, text, or formulas can be entered into a cell. When a spreadsheet is used for data entry the data from each subject is usually entered on a single row and each column usually contains a single variable measured on each subject. Excel, Quatro Pro, and Lotus 1-2-3 are currently popular spreadsheet programs. Database programs require users to build a data entry format by specifying the names and characteristics of the variables to be measured. The program will then prompt the user to enter the variables in the appropriate order. Users of spreadsheets can view all the data that has been entered while users of databases cannot. On the other hand, database programs allow users to build several different databases with related information and then to link these different databases.

The first consideration in selecting a data entry program is whether or not the statistical and graphical software can easily read the data entered through the entry program. All statistical and graphical programs are written with the anticipation that the data input to the program will be in a specified form. Most programs provide means for changing the format of the data; however, it is almost always easier to use a data entry program that gives the data in the optimal format for use by the statistical and graphical programs.

Graphical Software

The first thing done and the last thing done in any data analysis should be the construction of graphs and figures. The initial graphs will be done in order to examine the assumptions necessary for further data analysis and to provide insight into the data and the relationships among the variables. Typically these graphs will be numerous scatter plots, box and whisker plots, histograms, bar charts, and pie charts. Since the initial graphs are for the purpose of analysis they are usually done as quickly and easily as possible. The graphs that are done at the end of the analysis are for the purpose of presentation or publication. Obviously, these graphs need to be very polished and detailed. Graphics for publication usually need to be produced in a certain size. For these graphs it is necessary to have a program that gives a great deal of control and allows substantial control of the printed output.

Statistical Software

Statistical software is used to do the calculations required for the data analysis. There are a very large number of statistical software packages. These packages differ greatly in their capacities. The most complete packages such as

SAS, and SPSS still lack the ability to do much nonparametric analysis. This is important because much of the data in the medical/biological/health area should be analyzed using those methods. Moreover, it is difficult to produce certain graphics. Other statistical packages are more limited in their statistical capabilities but allow some graphics to be produced quite easily. The hallmark of good research is planning ahead. Before any data is collected all the steps in the analysis should be planned in detail. After this planning statistical software should be selected so that the software will allow the analysis to be completed easily.

Data Analysis Examples

At this time almost all statistical calculations are done using computer statistical packages. A computer package is a collection of programs that are related in some way. A statistical computer package is a collection of programs that do various statistical procedures. There are many different statistical packages. Included among the better known statistical packages are SAS, SPSS, BMDP, STATA, and STATISITCA. In order to use any statistical package you must do the following things:

1. Start the package
2. Tell the package where to find the data and how to read the data.
3. Identify the particular procedure to be used.
4. Specify which options, if any, the procedure should employ.

Each package will have its own specific way in which these specifications are to be made. This is less of a disadvantage than it may appear since if you know how to use one package it is usually quite easy to learn another package. For these examples the SAS system will be used. All of the examples will relate to the data immediately below. Because of the need to illustrate particular statistical procedures, some of the examples are somewhat artificial; however, each example represents a possible approach to each problem.

The examples all relate to the following experiment.

Researchers were interested in the benefits of intensive personal attention on the success of diabetics in controlling their blood sugar. The subjects were 53 individuals with type II diabetes. The subjects were selected because their physicians were not satisfied with the blood sugar control the patients were achieving. The 53 subjects were randomly assigned to one of four treatment groups. All of the subjects were given the same instructions regarding diet, exercise, and medication. The treatment groups differed only in the amount of personal attention the subjects received from their physician and from other health professionals. The subjects in Group I received the least attention while the subjects in Group IV received the most. Groups II and III were intermediate in the individual attention they received. The following measurements were recorded for each subject:

SUBJ   Subject ID Number
AGE   Age in Years at Start of the Study
SEX   Sex coded as 0 = Female 1 = Male
WTKG1   Weight in Kilograms at the Start of the Study
HTCM   Height in Centimeters
GROUP   Treatment Group coded as 1, 2, 3, 4
BG1   Fasting Blood Glucose in mg/dl At Baseline
BG2   Fasting Blood Glucose in mg/dl After 2 Months on Treatment
BG3   Fasting Blood Glucose in mg/dl After 4 Months on Treatment
BG4   Fasting Blood Glucose in mg/dl After 6 Months on Treatment
WTKG2   Weight in Kilograms at the End of the Study (6 months)

The data are in an ASCII file named EXPLDATA. This file is stored on the S: drive on the computer being used. The file EXPLDATA contains only measurements. The first 4 lines of data in the file are:

```
1   50   1   84  169   4  153  121  124  95  81
2   56   0   64  160   3  145  128  112  97  62
3   48   0   55  148   3  133  116  100  85  53
4   57   0   64  160   4  147  116  126  96  62
```

Example 1: Summary Statistics

Every statistical analysis should start by generating appropriate descriptive statistics. The program below generates all of the most common summary statistics as well as some statistics that are used only infrequently.

```
DATA ONE ;
INFILE 'S:\EXPLDATA' ;            <<Gives the location of the data
INPUT SUBJ  AGE  SEX  WTKG1
```

```
        HTCM  GROUP  BG1  BG2
        BG3   BG4  WTKG2;         <<Gives the variable names in order
HTM = HTCM/100 ;        <<Creates the variable htm = height in meters
BMI1 = WTKG1/(HTM*HTM);   <<Create BMI and difference variables
BMI2 = WTKG2/(HTM*HTM) ;
DELTAWT = WTKG2 - WTKG1 ;
DELTABMI = BMI2 - BMI1 ;
AGEGP = FLOOR(AGE/10)*10;   <<Groups the ages into 10-year groups
RUN;
PROC SORT; BY SEX;
PROC UNIVARIATE; BY SEX;    <<Generates statistics for each Sex
TITLE 'SUMMARY STATISTICS' ;
RUN;
```

****

The output from this program is

****

Summary Statistics

--SEX=0 ---

Univariate Procedure
Variable=AGE

Moments					Quantiles(Def=5)				
N	26	Sum	26	100%	Max	64	99%	64
Mean	55.46	Wgts	1442	75%	Q3	59	95%	64
Std Dev	5.49	Sum	30.17	50%	Med	57	90%	61
Skewness	-0.54	Variance	-0.42	25%	Q1	53	10%	48
USS	80730	Kurtosis	754.46	0%	Min	44	5%	45
CV	9.90	CSS	1.07				1%	44
T:Mean=0	51.47	Std	0.0001					
Num^=0	26	Mean	26	Range	20			
M(Sign)	13	Pr>|T|	0.0001	Q3-Q1	6			
Sgn Rank	175.5	Num>0	0.0001	Mode	58			
		Pr>=|M|						
		Pr>=|S|						

Extremes

Lowest	Obs	Highest	Obs
44(	23)	61(	10)
45(	13)	61(	12)
48(	9)	61(	16)
48(	2)	64(	8)
49(	25)	64(	24)

Summary Statistics
-------------------------------------- SEX=0 --------------------------------------

Univariate Procedure

Variable=WTKG1

Moments				Quantiles(Def=5)				
N	26	Sum Wgts	26	100%	Max	68	99%	68
Mean	63.15	Sum	1642	75%	Q3	66	95%	67
Std Dev	4.10	Variance	16.85	50%	Med	64	90%	67
Skewness	-1.20	Kurtosis	0.59	25%	Q1	62	10%	55
USS	104120	CSS	421.38	0%	Min	53	5%	55
CV	500841	Std Mean	0.80				1%	53
T:Mean=0	78.43	Pr>|T|	0.0001					
Num^=0	26	Num>0	26	Range	15			
M(Sign)	13	Pr>=|M|	0.0001	Q3-Q1	4			
Sgn Rank	175.5	Pr>=|S|	0.0001	Mode	64			

Extremes

Lowest	Obs	Highest	Obs
53(	23)	67(	8)
55(	13)	67(	10)
55(	2)	67(	12)
58(	9)	67(	16)
59(	6)	68(	24)

Summary Statistics
-------------------------------------- SEX=0 --------------------------------------

Univariate Procedure
Variable=HTCM

Moments				Quantiles(Def=5)				
N	26	Sum Wgts	26	100%	Max	166	99%	166
Mean	159.11	Sum	4137	75%	Q3	163	95%	165
Std Dev	5.79	Variance	33.62	50%	Med	160	90%	165
Skewness	-1.28	Kurtosis	1.09	25%	Q1	158	10%	148
USS	659101	CSS	840.65	0%	Min	144	5%	147
CV	3.64	Std Mean	1.13				1%	144
T:Mean=0	139.9	Pr>|T|	0.0001					
Num^=0	26	Num>0	26	Range	22			
M(Sign)	13	Pr>=|M|	0.0001	Q3-Q1	5			
Sgn Rank	175.5	Pr>=|S|	0.0001	Mode	160			

Extremes

Lowest	Obs	Highest	Obs
144(	23)	164(	10)
147(	13)	165(	8)
148(	2)	165(	12)
152(	9)	165(	16)
154(	6)	166(	24)

Summary Statistics

-------------------------------------- SEX=0 --------------------------------------

Univariate Procedure
Variable=BG1

Moments					Quantiles(Def=5)				
N	26	Sum Wgts	26	100%	Max	159	99%	159
Mean	144.76	Sum	3674	75%	Q3	150	95%	159
Std Dev	8.89	Variance	79.14	50%	Med	147	90%	154
Skewness	-0.40	Kurtosis	-0.68	25%	Q1	136	10%	133
USS	54689	CSS	1978.6	0%	Min	128	5%	128
CV	6.14	Std Mean	1.74				1%	128
T:Mean=0	82.97	Pr>|T|	0.0001					
Num^=0	26	Num>0	26	Range	31			
M(Sign)	13	Pr>=|M|	0.0001	Q3-Q1	14			
Sgn Rank	175.5	Pr>=|S|	0.0001	Mode	149			

Extremes

Lowest	Obs	Highest	Obs
128(	23)	154(	10)
128(	13)	154(	12)
133(	9)	154(	16)
133(	2)	159(	8)
134(	6)	159(	24)

Summary Statistics
-- SEX=0 --

Univariate Procedure
Variable=BG2

Moments							Quantiles(Def=5)		
N	26	Sum Wgts	26	100%	Max	158	99%	158
Mean	130.96	Sum	3405	75%	Q3	142	95%	152
Std Dev	13.55	Variance	183.63	50%	Med	130	90%	148
Skewness	0.01	Kurtosis	-0.42	25%	Q1	120	10%	116
USS	45051	CSS	4590.9	0%	Min	102	5%	112
CV	10.34	Std Mean	2.65				1%	102
T:Mean=0	49.27	Pr>|T|	0.0001					
Num^=0	26	Num>0	26	Range	56			
M(Sign)	13	Pr>=|M|	0.0001	Q3-Q1	22			
Sgn Rank	175.5	Pr>=|S|	0.0001	Mode	116			

Extremes

Lowest	Obs	Highest	Obs
102(	9)	146(	5)
112(	22)	146(	12)
116(	3)	148(	20)
116(	2)	152(	10)
118(	15)	158(	24)

==
Similar output is obtained for the females for each of the variables BG3, BG4, WTKG2, HTM, BMI1, BMI2, DELTAWT, DELTABMI. These are omitted to save space. Male output is also omitted.
==

Example 2 A Simple Histogram

This example creates a simple histogram that is useful for the initial examination of the distribution of the data. Note that the portion of the program that gets the data ready for analysis is the same in all the examples. In this example histograms are generated only for the blood glucose variables BG1 and BG4 in order to save space.

```
DATA ONE ;
INFILE 'S:\EXPLDATA' ;                  <<Gives the location of the data
INPUT SUBJ AGE SEX WTKG1
```

```
        HTCM  GROUP  BG1  BG2
        BG3   BG4 WTKG2;          <<Gives the variable names in order
HTM = HTCM/100 ;           <<Creates the variable htm = height in meters
BMI1 = WTKG1/(HTM*HTM);    <<Create BMI and difference variables
BMI2 = WTKG2/(HTM*HTM) ;
DELTAWT = WTKG2 - WTKG1 ;
DELTABMI = BMI2 - BMI1 ;
AGEGP = FLOOR(AGE/10)*10;   <<Groups the ages into 10-year groups
RUN;
PROC CHART;
HBAR BG1 BG4 ;
TITLE 'COMPUTER HISTOGRAM' ;
RUN;
```

****

The output from this program is

****

Computer Histogram
  BG1

Midpoint		Freq	Cum. Freq	Percent	Cum. Percent
132	|******	6	6	11.54	11.54
140	|******	6	12	11.54	23.08
148	|**************	15	27	28.85	51.92
156	|************	13	40	25.00	76.92
164	|*********	9	49	17.31	94.23
172	|***	3	52	5.77	100.00

```
        -----------------------------------
        2   4   6   8   10   12   14
                   Frequency
```

Computer Histogram
BG4

Midpoint		Freq	Cum. Freq	Percent	Cum. Percent
75	|***	3	3	5.77	5.77
90	|*************	14	17	26.92	32.69
105	|********	8	25	15.38	48.08
120	|******	7	32	13.46	61.54
135	|******	7	39	13.46	75.00
150	|**********	10	49	19.23	94.23
165	|***	3	52	5.77	100.00

```
          ----------------------------------------------
          1  2  3  4  5  6  7  8  9  10  11  12  13  14
```

Frequency

===

Note that even though the two histograms were created at the same time they have different scales.

===

EXAMPLE 3 A t-test for independent groups

This example tests whether the ages and BG! Measurements differ between the males and the females without regard to group.

```
DATA ONE ;
INFILE 'S:\EXPLDATA' ;              <<Gives the location of the data
INPUT  SUBJ  AGE SEX WTKG1
        HTCM  GROUP  BG1  BG2
        BG3    BG4  WTKG2;          <<Gives the variable names in order
HTM = HTCM/100 ;          <<Creates the variable htm = height in meters
BMI1 = WTKG1/(HTM*HTM);   <<Create BMI and difference variables
BMI2 = WTKG2/(HTM*HTM) ;
DELTAWT = WTKG2 - WTKG1 ;
DELTABMI = BMI2 - BMI1 ;
AGEGP = FLOOR(AGE/10)*10;   <<Groups the ages into 10-year groups
RUN;
PROC TTEST;
CLASS SEX ;
```

```
VAR AGE BG1 ;
TITLE 'T TEST FOR INDEPENDENT GROUPS' ;
RUN;
```

****

The output for this program is

****

T Test for Independent Groups

TTEST PROCEDURE

Variable: AGE

SEX	N	Mean	Std Dev	Std Error	Variances	T	DF	Prob>|T|
0	26	55.46	5.49	1.07	Unequal	1.17	50	0.24
1	26	53.69	5.40	1.06	Equal	1.17	50	0.24

For H0: Variances are equal,

$F' = 1.03$ $DF = (25,25)$ $Prob>F' = 0.9391$

****

Variable: BG1

SEX	N	Mean	Std Dev	Std Error	Variances	T	DF	Prob>|T|
0	26	144.76	8.89	1.74	Unequal	-4.71	48.8	0.0001
1	26	157.42	10.38	2.03	Equal	-4.71	50.0	0.0000

For H0: Variances are equal,

$F' = 1.36$ $DF = (25,25)$ $Prob>F' = 0.4435$

===

In this case there is not a significant difference between the ages but the BG1 measurements for the males are significantly higher among the males than the females.

===

EXAMPLE 4 Analysis of a Contingency Table

This program generates three contingency tables and calculates a chi-square for each. The command "EXACT" which follows CHISQ requests that the Fisher-Irwin test be run in case the expected values are small.

```
DATA ONE ;
INFILE 'S:\EXPLDATA' ;                    <<Gives the location of the data
INPUT  SUBJ  AGE  SEX  WTKG1
       HTCM  GROUP  BG1  BG2
       BG3   BG4  WTKG2;               <<Gives the variable names in order
HTM = HTCM/100 ;            <<Creates the variable htm = height in meters
BMI1 = WTKG1/(HTM*HTM);      <<Create BMI and difference variables
BMI2 = WTKG2/(HTM*HTM) ;
DELTAWT = WTKG2 - WTKG1 ;
DELTABMI = BMI2 - BMI1 ;
AGEGP = FLOOR(AGE/10)*10;    <<Groups the ages into 10-year groups
RUN;
PROC FREQ;
TABLES  SEX*AGEGP  SEX*GROUP  AGEGP*GROUP
     / CHISQ EXACT ;
TITLE 'CHI-SQUARE TEST ON A CONTINGENCY TABLE';
RUN;
```

****
The output from this program is
****

Chi-Square Test on a Contingency Table

      TABLE OF SEX BY AGEGP

SEX   AGEGP

```
Frequency |
Percent   |
Row Pct   |
Col Pct   |      40|     50|     60| Total
-----------------------------------------------
        0 |     6 |    14|     6|    26
          | 11.54 | 26.92 | 11.54 |    50.00
          | 23.08 | 53.85 | 23.08 |
```

| | 50.00 | 46.67 | 60.00 | |

| 1 | 6 | 16 | 4 | 26
| | 11.54 | 30.77 | 7.69 | 50.00
| | 23.08 | 61.54 | 15.38 |
| | 50.00 | 53.33 | 40.00 |

Total   12   30   10   52
             23.08   57.69   19.23 100.00

STATISTICS FOR TABLE OF SEX BY AGEGP

Statistic	DF	Value	Prob
Chi-Square	2	0.533	0.766
Likelihood Ratio Chi-Square	2	0.536	0.765
Mantel-Haenszel Chi-Square	1	0.179	0.672
Fisher's Exact Test (2-Tail)			0.810
Phi Coefficient		0.101	
Contingency Coefficient		0.101	
Cramer's V		0.101	

Sample Size = 52

Chi-Square Test on a Contingency Table

TABLE OF SEX BY GROUP

SEX   GROUP

Frequency |
Percent |
Row Pct |
Col Pct | 1| 2| 3| 4| Total
----
      0 | 7| 6| 7| 6| 26
        | 13.46| 11.54| 13.46| 11.54| 50.00
        | 26.92| 23.08| 26.92| 23.08|
        | 53.85| 46.15| 53.85| 46.15|
----
      1 | 6| 7| 6| 7| 26

| 11.54| 13.46| 11.54| 13.46| 50.00
| 23.08| 26.92| 23.08| 26.92|
| 46.15| 53.85| 46.15| 53.85|

```
---------------------------------------------
```

| Total | 13 | 13 | 13 | 13 | 52 |
| | 25.00 | 25.00 | 25.00 | 25.00 | 100.00 |

STATISTICS FOR TABLE OF SEX BY GROUP

Statistic	DF	Value	Prob
Chi-Square	3	0.308	0.959
Likelihood Ratio Chi-Square	3	0.308	0.959
Mantel-Haenszel Chi-Square	1	0.060	0.806
Fisher's Exact Test (2-Tail)			1.000
Phi Coefficient		0.077	
Contingency Coefficient		0.077	
Cramer's V		0.077	

Sample Size = 52

Chi-Square Test on a Contingency Table

TABLE OF AGEGP BY GROUP
AGEGP   GROUP
Frequency |
Percent |
Row Pct |

Col Pct	1|	2|	3|	4|	Total

40	3		4		3		2		12
	5.77		7.69		5.77		3.85		23.08
	25.00		33.33		25.00		16.67		
	23.08		30.77		23.08		15.38		

```
---------------------------------------------
```

50	7		7		8		8		30
	13.46		13.46		15.38		15.38		57.69
	23.33		23.33		26.67		26.67		
	53.85		53.85		61.54		61.54		

```
---------------------------------------------
```

60	| 3|	2|	2|	3|	10
	| 5.77|	3.85|	3.85|	5.77|	19.23
	| 30.00|	20.00|	20.00|	30.00|	
	| 23.08|	15.38|	15.38|	23.08|	

```
----------------------------------------------
```

Total	13	13	13	13	52
	25.00	25.00	25.00	25.00	100.00

STATISTICS FOR TABLE OF AGEGP BY GROUP

Statistic	DF	Value	Prob
Chi-Square	6	1.200	0.977
Likelihood Ratio Chi-Square	6	1.216	0.976
Mantel-Haenszel Chi-Square	1	0.143	0.705
Fisher's Exact Test (2-Tail)			0.993
Phi Coefficient		0.152	
Contingency Coefficient		0.150	
Cramer's V		0.107	

Sample Size = 52
WARNING: 67% of cells have expected counts less
than 5. Chi-Square may not be a valid test.

EXAMPLE 5 Calculation of correlation coefficients.

This program calculates both the Pearson and the Spearman correlations
among the blood glucose measurements. The first part of the program
calculates the correlations without regard to group. The second part of the
program calculates the correlations within each group.

```
DATA ONE ;
INFILE 'S:\EXPLDATA' ;              <<Gives the location of the data
INPUT  SUBJ  AGE  SEX  WTKG1
         HTCM  GROUP  BG1  BG2
         BG3   BG4  WTKG2;          <<Gives the variable names in order
HTM = HTCM/100 ;         <<Creates the variable htm = height in meters
BMI1 = WTKG1/(HTM*HTM);  <<Create BMI and difference variables
BMI2 = WTKG2/(HTM*HTM) ;
```

```
DELTAWT = WTKG2 - WTKG1 ;
DELTABMI = BMI2 - BMI1 ;
AGEGP = FLOOR(AGE/10)*10;    <<Groups the ages into 10-year groups
RUN;
PROC CORR PEARSON SPEARMAN ;
VAR BG1 BG2 BG3 BG4 ;
TITLE 'OVERALL CORRELATIONS';

PROC SORT; BY GROUP;
PROC CORR PEARSON SPEARMAN; BY GROUP;
VAR BG1 BG2 BG3 BG4 ;
TITLE 'CORRELATIONS BY GROUP';
RUN;
```

****

The output from this program is

****

Overall Correlations

Correlation Analysis

4 'VAR' Variables: BG1   BG2   BG3   BG4

Simple Statistics

Variable	N	Mean	Std Dev	Median	Minimum	Maximum
BG1	52	151.09	11.51	150.50	128.00	175.00
BG2	52	137.17	15.53	135.50	102.00	171.00
BG3	52	129.90	17.33	126.50	100.00	170.00
BG4	52	117.19	26.17	115.50	78.00	168.00

Pearson Correlation Coefficients / Prob > |R| under Ho: Rho=0 N = 52

	BG1	BG2	BG3	BG4
BG1	1.00000	0.66934	0.44771	0.25487
	0.0	0.0001	0.0009	0.0682
BG2	0.66934	1.00000	0.88159	0.85922
	0.0001	0.0	0.0001	0.0001
BG3	0.44771	0.88159	1.00000	0.95861

	0.0009	0.0001	0.0	0.0001
BG4	0.25487	0.85922	0.95861	1.00000
	0.0682	0.0001	0.0001	0.0

Spearman Correlation Coefficients / Prob > |R| under Ho: Rho=0 N = 52

	BG1	BG2	BG3	BG4
BG1	1.00000	0.65339	0.40974	0.25125
	0.0	0.0001	0.0026	0.0724
BG2	0.65339	1.00000	0.87018	0.85867
	0.0001	0.0	0.0001	0.0001
BG3	0.40974	0.87018	1.00000	0.94917
	0.0026	0.0001	0.0	0.0001
BG4	0.25125	0.85867	0.94917	1.00000
	0.0724	0.0001	0.0001	0.0

Correlations by Group
--------------------------------------GROUP=1 ---

Correlation Analysis

4 'VAR' Variables: BG1   BG2   BG3   BG4

Simple Statistics

Variable	N	Mean	Std Dev	Median	Minimum	Maximum
BG1	13	151.07	12.06	153.00	128.00	172.00
BG2	13	150.23	12.04	152.00	127.00	171.00
BG3	13	149.38	11.90	151.00	126.00	170.00
BG4	13	148.53	11.73	150.00	125.00	168.00

Pearson Correlation Coefficients / Prob > |R| under Ho: Rho=0 N = 13

	BG1	BG2	BG3	BG4
BG1	1.00000	0.99894	0.99754	0.99354
	0.0	0.0001	0.0001	0.0001
BG2	0.99894	1.00000	0.99900	0.99658
	0.0001	0.0	0.0001	0.0001
BG3	0.99754	0.99900	1.00000	0.99840
	0.0001	0.0001	0.0	0.0001
BG4	0.99354	0.99658	0.99840	1.00000
	0.0001	0.0001	0.0001	0.0

Spearman Correlation Coefficients / Prob > |R| under Ho: Rho=0 N = 13

	BG1	BG2	BG3	BG4
BG1	1.00000	0.99312	0.99312	0.98487
	0.0	0.0001	0.0001	0.0001
BG2	0.99312	1.00000	1.00000	0.99451
	0.0001	0.0	0.0001	0.0001
BG3	0.99312	1.00000	1.00000	0.99451
	0.0001	0.0001	0.0	0.0001
BG4	0.98487	0.99451	0.99451	1.00000
	0.0001	0.0001	0.0001	0.0

===
Similar output from Groups 2, 3, and 4 is omitted to save space.
===

EXAMPLE 6 Simple Linear Regression

In this example the program calculates the regression line of BG1 on age. The program also tests the null hypothesis that the slope of the population regression line is zero.

```
DATA ONE ;
INFILE 'S:\EXPLDATA' ;                    <<Gives the location of the data
INPUT  SUBJ   AGE  SEX  WTKG1
        HTCM  GROUP  BG1  BG2
        BG3   BG4  WTKG2;         <<Gives the variable names in order
HTM = HTCM/100 ;         <<Creates the variable htm = height in meters
BMI1 = WTKG1/(HTM*HTM);    <<Create BMI and difference variables
BMI2 = WTKG2/(HTM*HTM) ;
DELTAWT = WTKG2 - WTKG1 ;
DELTABMI = BMI2 - BMI1 ;
AGEGP = FLOOR(AGE/10)*10;   <<Groups the ages into 10-year groups
RUN;
PROC GLM ;
MODEL BG1 = AGE  ;
TITLE 'SIMPLE REGRESSION' ;
RUN;
```

****
The output from this program is
****

Simple Regression
General Linear Models Procedure
Dependent Variable: BG1

Source	DF	Sum of Squares	Mean Square	F Value	Pr > F
Model	1	3384.47	3384.47	50.15	0.0001
Error	50	3374.04	67.48		
Total	51	6758.51			

R-Square	C.V.	Root MSE	BG1 Mean
0.500772	5.43	8.21	151.09

Source	DF	Sum of Squares	Mean Square	F Value	Pr > F
AGE	1	3384.47	3384.47	50.15	0.0001

Parameter	Estimate	T for H0: Parameter=0	Pr. > |T|	Std Error of Parameter Estimate
INTERCEPT	69.83	6.06	0.0001	11.53
AGE	1.48	7.08	0.0001	0.21

==
= Note that although the R-Square is only .500772 the line given by
$$BG1 = 69.83 + 1.48*AGE$$
has a slope that is significantly different from zero and thus the line gives a better prediction of the mean of BG1 for each AGE than the sample mean alone.
==

EXAMPLE 7 Multiple Regression

In this example we calculate a multiple regression equation to try to get a better regression equation by using age and BMI1 to predict BG1. This program also creates a SAS data set called GGG. This data set contains estimates of BG1 based on the regression equation. These estimates could be used in other calculations or in plots or graphs.

```
DATA ONE ;
INFILE 'S:\EXPLDATA' ;                    <<Gives the location of the data
INPUT  SUBJ  AGE SEX WTKG1
       HTCM  GROUP BG1 BG2
       BG3   BG4 WTKG2;          <<Gives the variable names in order
HTM = HTCM/100 ;          <<Creates the variable htm = height in meters
BMI1 = WTKG1/(HTM*HTM);     <<Create BMI and difference variables
BMI2 = WTKG2/(HTM*HTM) ;
DELTAWT = WTKG2 - WTKG1 ;
DELTABMI = BMI2 - BMI1 ;
AGEGP = FLOOR(AGE/10)*10;    <<Groups the ages into 10-year groups
RUN;
PROC GLM ;
MODEL BG1 = AGE BMI1 ;
```

```
OUTPUT OUT= GGG P=Y ;                    <<The variable Y is the
estimate of BG1 from the regression
TITLE 'MULTIPLE REGRESSION' ;
RUN;
```

****
The output from this program is
****

Multiple Regression
General Linear Models Procedure

Dependent Variable: BG1

Source	DF	Sum of Squares	Mean Square	F Value	Pr > F
Model	2	6738.91	3369.45	8423.80	0.0001
Error	49	19.59	0.399		
Total	51	6758.51			

R-Square	C.V.	Root MSE	BG1 Mean
0.997100	0.418575	0.63245003	151.09615385

Source	DF	Sum of Squares	Mean Square	F Value	Pr > F
AGE	1	4579.27	4579.27	11448.40	0.0001
BMI1	1	3354.44	3354.44	8386.25	0.0001

Parameter	Estimate	T for H0: Parameter=0	Pr > |T|	Std Error of Estimate
INTERCEPT	-51.59	-32.33	0.0001	1.5957296
AGE	1.76	107.00	0.0001	0.0164564
BMI1	39608.95	91.58	0.0001	432.5231181

===
Both AGE and BMI1 are significant. That is both are helpful in predicting BG1. This is reflected by the R-Square of .997100.
===

EXAMPLE 8 ONE WAY ANALYSIS OF VARIANCE WITHOUT REPEATED MEASURES.

In this example we test whether the population mean values for BG1 differ among the four treatment groups. The Duncan's New Multiple Range Test is used to examine the pattern of differences among the group means.

```
DATA ONE ;
INFILE 'S:\EXPLDATA' ;              <<Gives the location of the data
INPUT  SUBJ  AGE  SEX  WTKG1
        HTCM  GROUP  BG1  BG2
        BG3   BG4  WTKG2;          <<Gives the variable names in order
HTM = HTCM/100 ;           <<Creates the variable htm = height in meters
BMI1 = WTKG1/(HTM*HTM);     <<Create BMI and difference variables
BMI2 = WTKG2/(HTM*HTM) ;
DELTAWT = WTKG2 - WTKG1 ;
DELTABMI = BMI2 - BMI1 ;
AGEGP = FLOOR(AGE/10)*10;    <<Groups the ages into 10-year groups
RUN;
PROC GLM ;
CLASS GROUP;
MODEL BG1 = GROUP;
MEANS GROUP / DUNCAN ;
TITLE 'ONE-WAY ANALYSIS OF VARIANCE FOR INDEPENDENT
GROUPS';
RUN;
```

****
The output from this program is
****

One-Way Analysis of Variance for Independent Groups
General Linear Models Procedure

Dependent Variable: BG1

Source	DF	Sum of Squares	Mean Square	F Value	Pr > F
Model	3	132.67	44.22	0.32	0.8106
Error	48	6625.84	138.03		
Total	51	6758.51			

R-Square	C.V.	Root MSE	BG1 Mean
0.019630	7.775828	11.74897704	151.09615385

Source	DF	Sum of Squares	Mean Square	F Value	Pr > F
GROUP	3	132.67	44.22	0.32	0.8106

Duncan's Multiple Range Test for variable: BG1
Alpha= 0.05 df= 48 MSE= 138.0385

Means with the same letter are not significantly different.

Duncan Grouping	Mean	N	GROUP
A	153.538	13	4
A			
A	151.077	13	1
A			
A	150.692	13	2
A			
A	149.077	13	3

===

In this example the ANOVA does not detect a difference among the population means. The Duncan's test reflects this by putting the letter "A" beside each mean. If there had been significant differences the Duncan's procedure would have put different letters beside means of groups that differed significantly.

===

EXAMPLE 9 TWO FACTOR ANALYSIS OF VARIANCE FOR INDEPENDENT GROUPS

In this example we look at the eight populations defined by Treatment Group and Sex. The two factor analysis of variance tests whether the means of the four groups differ, whether the means of the two sexes differ, and whether there is any interaction. If we had desired, a multiple comparison procedure such as the Duncan's above could have been used.

```
DATA ONE ;
INFILE 'S:\EXPLDATA' ;                 <<Gives the location of the data
INPUT  SUBJ  AGE  SEX  WTKG1
       HTCM  GROUP  BG1  BG2
       BG3   BG4  WTKG2;         <<Gives the variable names in order
HTM = HTCM/100 ;         <<Creates the variable htm = height in meters
BMI1 = WTKG1/(HTM*HTM);    <<Create BMI and difference variables
BMI2 = WTKG2/(HTM*HTM) ;
DELTAWT = WTKG2 - WTKG1 ;
DELTABMI = BMI2 - BMI1 ;
AGEGP = FLOOR(AGE/10)*10;   <<Groups the ages into 10-year groups
RUN;
PROC GLM ;
CLASS GROUP SEX;
MODEL BG1 = GROUP SEX GROUP*SEX ;
TITLE 'TWO FACTOR ANALYSIS OF VARIANCE FOR INDEPENDENT
GROUPS';
RUN;
```

```
*******************************************
The output from this program is
*******************************************
```

Two Factor Analysis of Variance for Independent Groups
General Linear Models Procedure

Dependent Variable: BG1

Source	DF	Sum of Squares	Mean Square	F Value	Pr > F
Model	7	2195.42	313.63	3.02	0.0110

Error	44	4563.09	103.70670996	
Total	51	6758.51		

R-Square	C.V.	Root MSE	BG1 Mean
0.324838	6.73	10.18	151.09

Source	DF	Sum of Squares	Mean Square	F Value	Pr>F
GROUP	3	98.74	32.91	0.32	0.8127
SEX	1	2042.37	2042.37	19.69	0.0001
GROUP*SEX	3	20.37	6.79	0.07	0.9779

==
In this case we see that there is no group by sex interaction (GROUP*SEX) and that the group means are not significantly different. This is not surprising since the BG1 measurement is made prior to treatment. The ANOVA does show a statistically significant difference between the means for the males and the females.

==

EXAMPLE 10 SINGLE FACTOR ANALYSIS OF VARIANCE WITH REPEATED MEASURES

In this example we calculate four separate analyses of variance. One analysis within each treatment group. The null hypothesis being tested is that there is no difference among the means of the blood glucose measurements at the four times.

```
DATA ONE ;
INFILE 'S:\EXPLDATA' ;               <<Gives the location of the data
INPUT  SUBJ  AGE  SEX  WTKG1
       HTCM  GROUP  BG1  BG2
       BG3   BG4  WTKG2;            <<Gives the variable names in order
HTM = HTCM/100 ;         <<Creates the variable htm = height in meters
BMI1 = WTKG1/(HTM*HTM);    <<Create BMI and difference variables
BMI2 = WTKG2/(HTM*HTM) ;
DELTAWT = WTKG2 - WTKG1 ;
DELTABMI = BMI2 - BMI1 ;
AGEGP = FLOOR(AGE/10)*10;   <<Groups the ages into 10-year groups
RUN;
PROC SORT; BY GROUP;
```

```
RUN;
PROC SORT; BY GROUP;
PROC GLM ; BY GROUP  ;
CLASS GROUP  ;
MODEL BG1 BG2 BG3 BG4 = GROUP   ;
REPEATED TIME 4 ;
TITLE 'SINGLE FACTOR ANALYSIS OF VARIANCE WITH REPEATED
MEASURES';
RUN;
```

****

The output from this program is
****

NOTE: This program generates an extremely lengthy output. The parts that are not related to this example are omitted.

Single Factor Analysis of Variance with Repeated Measures

----------------------------------GROUP=1 ---

General Linear Models Procedure
Repeated Measures Analysis of Variance
Univariate Tests of Hypotheses for Within Subject Effects

Source: TIME

DF	Sum of Squares	Mean Square	F Value	Pr > F	Adj Pr > F G - G	Adj Pr > F H - F
3	46.53	15.51	38.62	0.0001	0.0001	0.0001

Source: Error(TIME)

DF	Type III SS	Mean Square
36	14.46153846	0.40170940

Greenhouse-Geisser Epsilon = 0.4349
Huynh-Feldt Epsilon = 0.4663

---------------------------------GROUP=2 --------------------------------------

General Linear Models Procedure
Repeated Measures Analysis of Variance
Univariate Tests of Hypotheses for Within Subject Effects

Source: TIME

DF	Sum of Squares	Mean Square	F Value	Pr > F	G - G	Adj Pr > F H - F
3	2761.59	920.53	1431.26	0.0001	0.0001	0.0001

Source: Error(TIME)

DF	Type III SS	Mean Square
36	23.15384615	0.64316239

---------------------------------GROUP=3 --------------------------------------

General Linear Models Procedure
Repeated Measures Analysis of Variance
Univariate Tests of Hypotheses for Within Subject Effects

Source: TIME

DF	Sum of Squares	Mean Square	F Value	Pr > F	G - G	Adj Pr > F H - F
3	15752.36	5250.78	1981.75	0.0001	0.0001	0.0001

Source: Error(TIME)

DF	Type III SS	Mean Square
36	95.38461538	2.64957265

Greenhouse-Geisser Epsilon = 0.3724
Huynh-Feldt Epsilon = 0.3836

General Linear Models Procedure
Repeated Measures Analysis of Variance
Univariate Tests of Hypotheses for Within Subject Effects

Source: TIME

DF	Sum of Squares	Mean Square	F Value	Pr > F	G - G	Adj Pr > F H - F
3	29211.75	9737.25	172.47	0.0001	0.0001	0.0001

Source: Error(TIME)

DF	Type III SS	Mean Square
36	2032.50000000	56.45833333

Greenhouse-Geisser Epsilon = 0.3352
Huynh-Feldt Epsilon = 0.3357

==
From this analysis we see that within each group there are highly significant differences among the population means at the four times measurements of blood glucose were made. Multiple comparison tests such as the Duncan's New Multiple Range Test (above) could have been used to examine the pattern of differences among the means. The Greenhouse-Geisser epsilon (G - G) and the Huynh-Feldt epsilon (H - F) are different ways that have been suggested for "adjusting" the p-value. Statisticians disagree about the value of these adjustments. In these examples the adjustments did not make any difference.
==

EXAMPLE 11 TWO FACTOR ANALYSIS OF VARIANCE WITH REPEATED MEASURES

In this example we generate a two factor analysis of variance that tests whether there are differences among the four groups, or differences among the blood glucose means at the four times measurements were made, or whether there was any interaction between the groups and time. In this case a significant interaction would mean that the pattern of means across time was not the same in the four groups.

```
DATA ONE ;
INFILE 'S:\EXPLDATA' ;                  <<Gives the location of the data
INPUT  SUBJ   AGE  SEX  WTKG1
        HTCM  GROUP  BG1  BG2
        BG3   BG4  WTKG2;         <<Gives the variable names in order
HTM = HTCM/100 ;        <<Creates the variable htm = height in meters
BMI1 = WTKG1/(HTM*HTM);    <<Create BMI and difference variables
BMI2 = WTKG2/(HTM*HTM) ;
DELTAWT = WTKG2 - WTKG1 ;
DELTABMI = BMI2 - BMI1 ;
AGEGP = FLOOR(AGE/10)*10;    <<Groups the ages into 10-year groups
RUN;
PROC GLM ;
CLASS GROUP  ;
MODEL BG1 BG2 BG3 BG4 = GROUP  ;
REPEATED TIME 4 ;
TITLE 'TWO FACTOR ANALYSIS OF VARIANCE WITH REPEATED
MEASURES';
RUN;
```

****
The output from this program is
****

Once again the unnecessary portions of the output have been omitted.

Two Factor Analysis of Variance with Repeated Measures

General Linear Models Procedure
Repeated Measures Analysis of Variance
Tests of Hypotheses for Between Subjects Effects

Source	DF	Type III SS	Mean Square	F Value	Pr > F
GROUP	3	30078.28	10026.09	23.37	0.0001
Error	48	20589.73	428.95		

General Linear Models Procedure
Repeated Measures Analysis of Variance
Univariate Tests of Hypotheses for Within Subject Effects
Source: TIME

DF	Sum of Squares	Mean Square	F Value	Pr > F	Adj Pr > F G - G	Adj Pr > F H - F
3	31279.20	10426.40	693.33	0.0001	0.0001	0.0001

Source: TIME*GROUP

DF	Sum of Squares	Mean Square	F Value	Pr > F	Adj Pr > F G - G	Adj Pr > F H - F
9	16493.04	1832.56	121.86	0.0001	0.0001	0.0001

Source: Error(TIME)

DF	Type III SS	Mean Square
14	2165.50000000	15.03819444

Greenhouse-Geisser Epsilon = 0.3405
Huynh-Feldt Epsilon = 0.3627

===

This test gives an F Value of 23.37 (p = .0001) for testing the differences among the means of the four treatment groups. It also gives an F of 693.33 for testing the differences among the blood glucose means at the four times. HOWEVER, the ANOVA also has a highly significant interaction (TIME*GROUP). Because the interaction is significant the apparent differences among the groups and times should not be interpreted from this procedure without further investigation. The usually recommended procedure is to generate the four within group analyses of the means across times (these were generated in EXAMPLE 10). In this case the interpretation should be made after careful examination of the 16 Group-Time means and the ANOVAs generated in both Example 10 and Example 11.

===

Summary

Computers are complex systems that include hardware, an operating system, and applications software. The planning of the research project should include planning the computer and software resources needed to complete the project. These resources must work together so that the project can be completed smoothly. If computers and software are to be purchased the statistical and graphical applications should be selected first. These programs should be selected so that they will easily produce the output that has been planned. The hardware and operating system should be selected so that the statistical, graphical, and other necessary applications software will run optimally.

References

Everitt B. S. and Der G. "A Handbook of Statistical Analyses Using SAS" Boca Raton, FL: Chapman & Hall/CRC, 1998.

Dolorio F. C., and Hardy K. A., "Quick Start to Data Analysis With SAS" Belmont, CA: Wadsworth Publishing Co., 1996.

Hatcher L. and Stepanski E. J., "A Step by Step Approach to Using the SAS System for Univariate and Multivariate Statistics" Cary NC: SAS Institute, Inc., 1994.

Scholtzhauer S. D. and Littell R. C., "SAS System for Elementary Statistical Analysis (2nd. Ed.)" Cary, NC: SAS Institute, Inc., 1997.

APPENDIX A

Statistical Tables

A.1   Table of Random Numbers

A.2   The Standard Normal Curve

A.3   Critical Values of Correlation Coefficients

A.4   Transformation of r to z

A.5   Critical Values of t

A.6   Critical Values of F

A.7   Critical Values of the Studentized Range
       Statistic

A.8   Critical Values of Chi-Square

A.9   Values of Spearman for the .05 and .01
       Levels of Significance

Table A.1   Table of Random Numbers

22 17 68 65 84	68 95 23 92 35	87 02 22 57 51	61 09 43 95 06	58 24 82 03 47
19 36 27 59 46	13 79 93 37 55	39 77 32 77 09	85 52 05 30 62	47 83 51 62 74
16 77 23 02 77	09 61 87 25 21	28 06 24 25 93	16 71 13 59 78	23 05 47 47 25
78 43 76 71 61	20 44 90 32 64	97 67 63 99 61	46 38 03 93 22	69 81 21 99 21
03 28 28 26 08	73 37 32 04 05	69 30 16 09 05	88 69 58 29 99	35 07 44 75 47
93 22 53 64 39	07 10 63 76 35	87 03 04 79 88	08 13 13 85 51	55 34 57 72 69
78 76 58 54 74	92 38 70 96 92	52 06 79 79 45	82 63 18 27 44	69 66 92 19 09
23 68 35 26 00	99 53 93 61 28	52 70 05 48 34	56 65 05 61 86	90 92 10 70 80
15 39 25 70 99	93 86 52 77 65	15 33 59 05 28	22 87 26 07 47	86 96 98 29 06
58 71 96 30 24	18 46 23 34 27	85 13 99 24 44	49 18 09 79 49	74 16 32 23 02
57 35 27 33 72	24 53 63 94 09	41 10 76 47 91	44 04 95 49 66	39 60 04 59 81
48 50 86 54 48	22 06 34 72 52	82 21 15 65 20	33 29 94 71 11	15 91 29 12 03
61 96 48 95 03	07 16 39 33 66	98 56 10 56 79	77 21 30 27 12	90 49 22 23 62
36 93 89 41 26	29 70 83 63 51	99 74 20 52 36	87 09 41 15 09	98 60 16 03 03
18 87 00 42 31	57 90 12 02 07	23 47 37 17 31	54 08 01 88 63	39 41 88 92 10
88 56 53 27 59	33 35 72 67 47	77 34 55 45 70	08 18 27 38 90	16 95 86 70 75
09 72 95 84 29	49 41 31 06 70	42 38 06 45 18	64 84 73 31 65	52 53 37 97 15
12 96 88 17 31	65 19 69 02 83	60 75 86 90 68	24 64 19 35 51	56 61 87 39 12
85 94 57 24 16	92 09 84 38 76	22 00 27 69 85	29 81 94 78 70	21 94 47 90 12
38 64 43 59 98	98 77 87 68 07	91 51 67 62 44	40 98 05 93 78	23 32 65 41 18
53 44 09 42 72	00 41 86 79 79	68 47 22 00 20	35 55 31 51 51	00 83 63 22 55
40 76 66 26 84	57 99 99 90 37	36 63 32 08 58	37 40 13 68 97	87 64 81 07 83
02 17 79 18 05	12 59 52 57 02	22 07 90 47 03	28 14 11 30 79	20 69 22 40 98
95 17 82 06 53	31 51 10 96 46	92 06 88 07 77	56 11 50 81 69	40 23 72 51 39
35 76 22 42 92	96 11 83 44 80	34 68 35 48 77	33 42 40 90 60	73 96 53 97 86
26 29 13 56 41	85 47 04 66 08	34 72 57 59 13	82 43 80 46 15	38 26 61 70 04
77 80 20 75 82	72 82 32 99 90	63 95 73 76 63	89 73 44 99 05	48 67 26 43 18
46 40 66 44 52	91 36 74 43 53	30 82 13 54 00	78 45 63 98 35	55 03 36 67 68
37 56 08 18 09	77 53 84 46 47	31 91 18 95 58	24 16 74 11 53	44 10 13 85 57
61 65 61 68 66	37 27 47 39 19	84 83 70 07 48	53 21 40 06 71	95 06 79 88 54
93 43 69 64 07	34 18 04 52 35	56 27 09 24 86	61 85 53 83 45	19 90 70 99 00
21 96 60 12 99	11 20 99 45 18	48 13 93 55 34	18 37 79 49 90	65 97 38 20 46
95 20 47 97 97	27 37 83 28 71	00 06 41 41 74	45 89 09 39 84	51 67 11 52 49
97 86 21 78 73	10 65 81 92 59	58 76 17 14 97	04 76 62 16 17	17 95 70 45 80
69 92 06 34 13	59 71 74 17 32	27 55 10 24 19	23 71 82 13 74	63 52 52 01 41
04 31 17 21 56	33 73 99 19 87	26 72 39 27 67	53 77 57 68 93	60 61 97 22 61
61 06 98 03 91	87 14 77 43 96	43 00 65 98 50	45 60 33 01 07	98 99 46 50 47
85 93 85 86 88	72 87 08 62 40	16 06 10 89 20	23 21 34 74 97	76 38 03 29 63
21 74 32 47 45	73 96 07 94 52	09 65 90 77 47	25 76 16 19 33	53 05 70 53 30
15 69 53 82 80	79 96 23 53 10	65 39 07 16 29	45 33 02 43 70	02 87 40 41 45
02 89 08 04 49	20 21 14 68 86	87 63 93 95 17	11 29 01 95 80	35 14 97 35 33
87 18 15 89 79	85 43 01 72 73	08 61 74 51 69	89 74 39 82 15	94 51 33 41 67
98 83 71 94 22	59 97 50 99 52	08 52 85 08 40	87 80 61 65 31	91 51 80 32 44
10 08 58 21 66	72 68 49 29 31	89 85 84 46 06	59 73 19 85 23	65 09 29 75 63
47 90 56 10 08	88 02 84 27 83	42 29 72 23 19	66 56 45 65 79	20 71 53 20 25
22 85 61 68 90	49 64 92 85 44	16 40 12 89 88	50 14 49 81 06	01 82 77 45 12
67 80 43 79 33	12 83 11 41 16	25 58 19 68 70	77 02 54 00 52	53 43 37 15 26
27 62 50 96 72	79 44 61 40 15	14 53 40 65 39	27 31 58 50 28	11 39 03 34 25
33 78 80 87 15	38 30 06 38 21	14 47 47 07 26	54 96 87 53 32	40 36 40 96 76
13 13 92 66 99	47 24 49 57 74	32 25 43 62 17	10 97 11 69 84	99 63 22 32 98

10 27 53 96 23	71 50 54 36 23	54 31 04 82 98	04 14 12 15 09	26 78 25 47 47
28 41 50 61 88	64 85 27 20 18	83 36 36 05 56	39 71 65 09 62	94 76 62 11 89
34 21 42 57 02	59 19 18 97 48	80 30 03 30 98	05 24 67 70 07	84 97 50 87 46
61 81 77 23 23	82 82 11 54 08	53 28 70 58 96	44 07 39 55 43	42 34 43 39 28
61 15 18 13 54	16 86 20 26 88	90 74 80 55 09	14 53 90 51 17	52 01 63 01 59
91 76 21 64 64	44 91 13 32 97	75 31 62 66 54	84 80 32 75 77	56 08 25 70 29
00 97 79 08 06	37 30 28 59 85	53 56 68 53 40	01 74 39 59 73	30 19 99 85 48
36 46 18 34 94	75 20 80 27 77	78 91 69 16 00	08 43 18 73 68	67 69 61 34 25
88 98 99 60 50	65 95 79 42 94	93 62 40 89 96	43 56 47 71 66	46 76 29 67 02
04 37 59 87 21	05 02 03 24 17	47 97 81 56 51	92 34 86 01 82	55 51 33 12 91
63 62 06 34 41	94 21 78 55 09	72 76 45 16 94	29 95 81 83 83	79 88 01 97 30
78 47 23 53 90	34 41 92 45 71	09 23 70 70 07	12 38 92 79 43	14 85 11 47 23
87 68 62 15 43	53 14 36 59 25	54 47 33 70 15	59 24 48 40 35	50 03 42 99 36
47 60 92 10 77	88 59 53 11 52	66 25 69 07 04	48 68 64 71 06	61 65 70 22 12
56 88 87 59 41	65 28 04 67 53	95 79 88 37 31	50 41 06 94 76	81 83 17 16 33
02 57 45 86 67	73 43 07 34 48	44 26 87 93 29	77 09 61 67 84	06 69 44 77 75
31 54 14 13 17	48 62 11 90 60	68 12 93 64 28	46 24 79 16 76	14 60 25 51 01
28 50 16 43 36	28 97 85 58 99	67 22 52 76 23	24 70 36 54 54	59 28 61 71 96
63 29 62 66 50	02 63 45 52 38	67 63 47 54 75	83 24 78 43 20	92 63 13 47 48
45 65 58 26 51	76 96 59 38 72	86 57 45 71 46	44 67 76 14 55	44 88 01 62 12
39 65 36 63 70	77 45 85 50 51	74 13 39 35 22	30 53 36 02 95	49 34 88 73 61
73 71 98 16 04	29 18 94 51 23	76 51 94 84 86	79 93 96 38 63	08 58 25 58 94
72 20 56 20 11	72 65 71 08 86	79 57 95 13 91	97 48 72 66 48	09 71 17 24 89
75 17 26 99 76	89 37 20 70 01	77 31 61 95 46	26 97 05 73 51	53 33 18 72 87
37 48 60 82 29	81 30 15 39 14	48 38 75 93 29	06 87 37 78 48	45 56 00 84 47
68 08 02 80 72	83 71 46 30 49	89 17 95 88 29	02 39 56 03 46	97 74 06 56 17
14 23 98 61 67	70 52 85 01 50	01 84 02 78 43	10 62 98 19 41	18 83 99 47 99
49 08 96 21 44	25 27 99 41 28	07 41 08 34 66	19 42 74 39 91	41 96 53 78 72
78 37 06 08 43	63 61 62 42 29	39 68 95 10 96	09 24 23 00 62	56 12 80 73 16
37 21 34 17 68	68 96 83 23 56	32 84 60 15 31	44 73 67 34 77	91 15 79 74 58
14 29 09 34 04	87 83 07 55 07	76 58 30 83 64	87 29 25 58 84	86 50 60 00 25
58 43 28 06 36	49 52 83 51 14	47 56 91 29 34	05 87 31 06 95	12 45 57 09 09
10 43 67 29 70	80 62 80 03 42	10 80 21 38 84	90 56 35 03 09	43 12 74 49 14
44 38 88 39 54	86 97 37 44 22	00 95 01 31 76	17 16 29 56 63	38 78 94 49 81
90 69 59 19 51	85 39 52 85 13	07 28 37 07 61	11 16 36 27 03	78 86 72 04 95
41 47 10 25 62	97 05 31 03 61	20 26 36 31 62	68 69 86 95 44	84 95 48 46 45
91 94 14 63 19	75 89 11 47 11	31 56 34 19 09	79 57 92 36 59	14 93 87 81 40
80 06 54 18 66	09 18 94 06 19	98 40 07 17 81	22 45 44 84 11	24 62 20 42 31
67 72 77 63 48	84 08 31 55 58	24 33 45 77 58	80 45 67 93 82	75 70 16 08 24
59 40 24 13 27	79 26 88 86 30	01 31 60 10 39	53 58 47 70 93	85 81 56 39 38
05 90 35 89 95	01 61 16 96 94	50 78 13 69 36	37 68 53 37 31	71 26 35 03 71
44 43 80 69 98	46 68 05 14 82	90 78 50 05 62	77 79 13 57 44	59 60 10 39 66
61 81 31 96 82	00 57 25 60 59	46 72 60 18 77	55 66 12 62 11	08 99 55 64 57
42 88 07 10 05	24 98 65 63 21	47 21 61 88 32	27 80 30 21 60	10 92 35 36 12
77 94 30 05 39	28 10 99 00 27	12 73 73 99 12	49 99 57 94 82	96 88 57 17 91
78 83 19 76 16	94 11 68 84 26	23 54 20 86 85	23 86 66 99 07	36 37 34 92 09
87 76 59 61 81	43 63 64 61 61	65 76 36 95 90	18 48 27 45 68	27 23 65 30 72
91 43 05 96 47	55 78 99 95 24	37 55 85 78 78	01 48 41 19 10	35 19 54 07 73
84 97 77 72 73	09 62 06 65 72	87 12 49 03 60	41 15 20 76 27	50 47 02 29 16
87 41 60 76 83	44 88 96 07 80	85 05 83 38 96	73 70 66 81 90	30 56 10 48 59

Table A.1 is taken from Table XXXIII of Fisher, *Statistical Methods for Research Workers*, published by Oliver and Boyd, Ltd., Edinburgh.

Table A.2   The Standard Normal Curve

z	One tail π beyond	One tail π remainder	Two tail π beyond	Two tail π remainder	z	One tail π beyond	One tail π remainder	Two tail π beyond	Two tail π remainder
0.00	0.5000	0.5000	1.0000	0.0000	0.45	0.3264	0.6736	0.6527	0.3473
0.01	0.4960	0.5040	0.9920	0.0080	0.46	0.3228	0.6772	0.6455	0.3545
0.02	0.4920	0.5080	0.9840	0.0160	0.47	0.3192	0.6808	0.6384	0.3616
0.03	0.4880	0.5120	0.9761	0.0239	0.48	0.3156	0.6844	0.6312	0.3688
0.04	0.4840	0.5160	0.9681	0.0319	0.49	0.3121	0.6879	0.6241	0.3759
0.05	0.4801	0.7199	0.9601	0.0399	0.50	0.3085	0.6915	0.6171	0.3829
0.06	0.4761	0.5239	0.9522	0.0478	0.51	0.3050	0.6950	0.6101	0.3899
0.07	0.4721	0.5279	0.9442	0.0558	0.52	0.3015	0.8985	0.6031	0.3969
0.08	0.4681	0.5319	0.9362	0.0638	0.53	0.2981	0.7019	0.5961	0.4039
0.09	0.4641	0.5359	0.9283	0.0717	0.54	0.2946	0.7054	0.5892	0.4108
0.10	0.4602	0.5398	0.9203	0.0797	0.55	0.2912	0.7088	0.5823	0.4177
0.11	0.4562	0.5438	0.9124	0.0876	0.56	0.2877	0.7123	0.5755	0.4245
0.12	0.4522	0.4378	0.9045	0.0955	0.57	0.2843	0.7157	0.5687	0.4313
0.13	0.4483	0.5517	0.8966	0.1034	0.58	0.2810	0.7190	0.5619	0.4381
0.14	0.4443	0.5557	0.8887	0.1113	0.59	0.2776	0.7224	0.5552	0.4448
0.15	0.4404	0.5596	0.8808	0.1192	0.60	0.2743	0.7257	0.5485	0.4515
0.16	0.4364	0.5636	0.8729	0.1271	0.61	0.2709	0.7291	0.5419	0.4581
0.17	0.4325	0.5675	0.8650	0.1350	0.62	0.2676	0.7324	0.5353	0.4647
0.18	0.4286	0.5714	0.8493	0.1507	0.63	0.2643	0.7357	0.5276	0.4713
0.19	0.4247	0.5753	0.8493	0.1507	0.64	0.2611	0.7389	0.5222	0.4778
0.20	0.4207	0.5793	0.8415	0.1585	0.65	0.2578	0.7422	0.5157	0.4843
0.21	0.4168	0.5832	0.8337	0.1663	0.66	0.2546	0.7454	0.5093	0.4907
0.22	0.4129	0.5871	0.8259	0.1741	0.67	0.2514	0.7486	0.5029	0.4971
0.23	0.4090	0.5910	0.8181	0.1819	0.6745	0.25	0.75	0.50	0.50
0.24	0.4052	0.5948	0.8103	0.1897	0.68	0.2483	0.7517	0.4965	0.5035
0.25	0.4013	0.5987	0.8026	0.1974	0.69	0.2451	0.7549	0.4902	0.5098
0.26	0.3974	0.6026	0.7949	0.2051	0.70	0.2420	0.7580	0.4839	0.5161
0.27	0.3936	0.6064	0.7872	0.2128	0.71	0.2389	0.7611	0.4777	0.5223
0.28	0.2897	0.6103	0.7795	0.2205	0.72	0.2358	0.7642	0.4715	0.5285
0.29	0.3859	0.6141	0.7718	0.2282	0.73	0.2327	0.7673	0.4654	0.5346
0.30	0.3821	0.6179	0.7642	0.2358	0.74	0.2296	0.7704	0.4593	0.5407
0.31	0.3783	0.6217	0.7566	0.2434	0.75	0.2266	0.7734	0.4533	0.5467
0.32	0.3745	0.6255	0.7490	0.2510	0.76	0.2236	0.7764	0.4473	0.5527
0.33	0.3707	0.6293	0.7414	0.2586	0.77	0.2206	0.7794	0.4413	0.5587
0.34	0.3669	0.6331	0.7339	0.2661	0.78	0.2177	0.7823	0.4354	0.5646
0.35	0.3632	0.6368	0.7263	0.2737	0.79	0.2148	0.7852	0.4295	0.5705
0.36	0.3594	0.6406	0.7188	0.2812	0.80	0.2119	0.7881	0.4237	0.5763
0.37	0.3557	0.6443	0.7114	0.2886	0.81	0.2090	0.7910	0.4179	0.5821
0.38	0.3520	0.6480	0.7039	0.2961	0.82	0.2061	0.7939	0.4122	0.5878
0.39	0.3483	0.6517	0.6965	0.3035	0.83	0.2033	0.7967	0.4065	0.5935
0.40	0.3446	0.6554	0.6892	0.3108	0.84	0.2005	0.7995	0.4009	0.5991
0.41	0.3409	0.6591	0.6818	0.3182	0.8416	0.20	0.80	0.40	0.60
0.42	0.3372	0.6628	0.6745	0.3255	0.85	0.1997	0.8023	0.3953	0.6047
0.43	0.3336	0.6664	0.6672	0.3328	0.86	0.1949	0.8051	0.3898	0.6102
0.44	0.3300	0.6700	0.6599	0.3401	0.87	0.1922	0.8078	0.3843	0.6157

	One tail		Two tail			One tail		Two tail	
z	π beyond	π remainder	π beyond	π remainder	z	π beyond	π remainder	π beyond	π remainder
0.88	0.1894	0.8106	0.3789	0.6211	1.32	0.0934	0.9066	0.1868	0.8132
0.89	0.1867	0.8133	0.3735	0.6265	1.33	0.0918	0.9082	0.1835	0.8165
0.90	0.1841	0.8159	0.3681	0.6319	1.34	0.0901	0.9099	0.1802	0.8198
0.91	0.1814	0.8186	0.3628	0.6372	1.35	0.0885	0.9115	0.1770	0.8230
0.92	0.1788	0.8212	0.3576	0.6424	1.36	0.0869	0.9131	0.1738	0.8202
0.93	0.1762	0.8238	0.3524	0.6476	1.37	0.0853	0.9147	0.1707	0.8293
0.94	0.1736	0.8264	0.3472	0.6528	1.38	0.0838	0.9162	0.1676	0.8324
0.95	0.1711	0.8289	0.3421	0.6579	1.39	0.0823	0.9177	0.1645	0.8355
0.96	0.1685	0.8315	0.3371	0.6629	1.40	0.0808	0.9192	0.1615	0.8385
0.97	0.1660	0.8340	0.3320	0.6680	1.41	0.0793	0.9207	0.1585	0.8415
0.98	0.1635	0.8365	0.3271	0.6729	1.42	0.0778	0.9222	0.1556	0.8444
0.99	0.1611	0.8389	0.3222	0.6778	1.43	0.0764	0.9286	0.1527	0.8473
1.00	0.1587	0.8413	0.3173	0.6827	1.44	0.0749	0.9251	0.1499	0.8501
1.01	0.1562	0.8438	0.3125	0.6875	1.45	0.0735	0.9265	0.1471	0.8529
1.02	0.1539	0.8461	0.3077	0.6923	1.46	0.0721	0.9279	0.1443	0.8567
1.03	0.1515	0.8485	0.3030	0.6970	1.47	0.0708	0.9292	0.1416	0.8584
1.04	0.1492	0.8508	0.2983	0.7017	1.48	0.0694	0.9306	0.1389	0.8611
1.05	0.1469	0.8531	0.2937	0.7063	1.49	0.0681	0.9319	0.1362	0.8638
1.06	0.1446	0.8554	0.2891	0.7109	1.50	0.0668	0.9332	0.1336	0.8664
1.07	0.1423	0.8577	0.2846	0.7154	1.51	0.0655	0.9345	0.1310	0.8690
1.08	0.1401	0.8599	0.2801	0.7199	1.52	0.0643	0.9357	0.1285	0.8715
1.09	0.1379	0.6621	0.2757	0.7243	1.53	0.0630	0.9370	0.1260	0.8740
1.10	0.1357	0.8643	0.2713	0.7287	1.54	0.0618	0.9382	0.1236	0.8764
1.11	0.1335	0.8665	0.2670	0.7330	1.55	0.0606	0.9394	0.1211	0.8789
1.12	0.1314	0.8686	0.2627	0.7373	1.56	0.0594	0.9406	0.1188	0.8812
1.13	0.1292	0.8708	0.2585	0.7415	1.57	0.0582	0.9418	0.1164	0.8836
1.14	0.1271	0.8729	0.2543	0.7457	1.58	0.0571	0.9429	0.1141	0.8859
1.15	0.1251	0.8749	0.2501	0.7499	1.59	0.0559	0.9441	0.1118	0.8882
1.16	0.1230	0.8770	0.2460	0.7540	1.60	0.0548	0.9452	0.1096	0.8904
1.17	0.1210	0.8790	0.2420	0.7580	1.61	0.0537	0.9463	0.1074	0.8926
1.18	0.1190	0.8810	0.2380	0.7620	1.62	0.0526	0.9474	0.1052	0.8948
1.19	0.1170	0.8830	0.2340	0.7660	1.63	0.0516	0.9484	0.1031	0.8969
1.20	0.1151	0.8049	0.2301	0.7699	1.64	0.0505	0.9495	0.1010	0.8990
1.21	0.1131	0.8869	0.2263	0.7737	1.645	0.05	0.95	0.10	0.90
1.22	0.1112	0.8888	0.2225	0.7775	1.65	0.0495	0.9505	0.0989	0.9011
1.23	0.1093	0.8907	0.2187	0.7813	1.66	0.0485	0.9515	0.0969	0.9031
1.24	0.1075	0.8925	0.2150	0.7890	1.67	0.0475	0.9525	0.0949	0.9051
1.25	0.1056	0.8944	0.2113	0.7887	1.68	0.0465	0.9535	0.0930	0.9070
1.26	0.1038	0.8962	0.2077	0.7923	1.69	0.0455	0.9545	0.0910	0.9090
1.27	0.1020	0.8980	0.2041	0.7959	1.70	0.0446	0.9554	0.0891	0.9109
1.28	0.1003	0.8997	0.2005	0.7995	1.71	0.0436	0.9564	0.0873	0.9127
1.282	0.10	0.90	0.20	0.80	1.72	0.0427	0.9573	0.0854	0.9146
1.29	0.0985	0.9015	0.1971	0.8029	1.73	0.0418	0.9582	0.0836	0.9164
1.30	0.0968	0.9032	0.1936	0.8064	1.74	0.0409	0.9591	0.0819	0.9181
1.31	0.0951	0.9049	0.1902	0.8098	1.75	0.0401	0.9599	0.0801	0.9199

(continued)

Table A.2 *(continued)*

z	One tail π beyond	One tail π remainder	Two tail π beyond	Two tail π remainder	z	One tail π beyond	One tail π remainder	Two tail π beyond	Two tail π remainder
1.76	0.0392	0.9608	0.0784	0.9216	2.20	0.0139	0.9661	0.0278	0.9722
1.77	0.0384	0.9616	0.0767	0.9233	2.21	0.0136	0.9864	0.0271	0.9729
1.78	0.0375	0.9625	0.0751	0.9249	2.22	0.0132	0.9868	0.0264	0.9736
1.79	0.0367	0.9633	0.0734	0.9266	2.23	0.0129	0.9871	0.0257	0.9743
1.80	0.0359	0.9641	0.0719	0.9281	2.24	0.0125	0.9875	0.0251	0.9749
1.81	0.0352	0.9649	0.0703	0.9297	2.25	0.0122	0.9878	0.0244	0.9756
1.82	0.0344	0.9656	0.0688	0.9312	2.26	0.0119	0.9881	0.0238	0.9762
1.83	0.0336	0.9664	0.0672	0.9328	2.27	0.0116	0.9884	0.0232	0.9768
1.84	0.0329	0.9671	0.0658	0.9342	2.28	0.0113	0.9887	0.0226	0.9774
1.85	0.0322	0.9678	0.0643	0.9357	2.29	0.0110	0.9890	0.0220	0.9780
1.86	0.0314	0.9686	0.0629	0.9371	2.30	0.0107	0.9893	0.0214	0.9786
1.87	0.0307	0.9693	0.0615	0.9385	2.31	0.0104	0.9896	0.0209	0.9791
1.88	0.0301	0.9699	0.0601	0.9399	2.32	0.0102	0.9898	0.0203	0.9797
1.89	0.0294	0.9706	0.0588	0.9412	2.326	0.01	0.99	0.02	0.98
1.90	0.0287	0.9713	0.0574	0.9426	2.33	0.0099	0.9901	0.0198	0.9802
1.91	0.0281	0.9719	0.0561	0.9439	2.34	0.0096	0.9904	0.0193	0.9807
1.92	0.0274	0.9726	0.0549	0.9451	2.35	0.0094	0.9906	0.0188	0.9812
1.93	0.0268	0.9732	0.0536	0.9464	2.36	0.0091	0.9909	0.0183	0.9817
1.94	0.0262	0.9738	0.0524	0.9476	2.37	0.0089	0.991	0.0178	0.9822
1.95	0.0256	0.9744	0.0512	0.9488	2.38	0.0087	0.9913	0.0173	0.9827
1.960	0.025	0.975	0.05	0.95	2.39	0.0084	0.9916	0.0168	0.9832
1.97	0.0244	0.9756	0.0488	0.9512	2.40	0.0082	0.9918	0.0164	0.9836
1.98	0.0239	0.9761	0.0477	0.9523	2.41	0.0080	0.9920	0.0160	0.9840
1.99	0.0233	0.9767	0.0466	0.9534	2.42	0.0078	0.9922	0.0155	0.9845
2.00	0.0228	0.9772	0.0455	0.9545	2.43	0.0075	0.9925	0.0151	0.9849
2.01	0.0222	0.9778	0.0444	0.9556	2.44	0.0073	0.9927	0.0147	0.9853
2.02	0.0217	0.9783	0.0434	0.9566	2.45	0.0071	0.9929	0.0143	0.9857
2.03	0.0212	0.9788	0.0424	0.9576	2.46	0.0069	0.9931	0.0139	0.9861
2.04	0.0207	0.9793	0.0414	0.9586	2.47	0.0068	0.9932	0.0135	0.9865
2.05	0.0202	0.9798	0.0404	0.9596	2.48	0.0066	0.9934	0.0131	0.9869
2.054	0.02	0.98	0.04	0.96	2.49	0.0064	0.9936	0.0128	0.9872
2.06	0.0197	0.9803	0.0394	0.9606	2.50	0.0062	0.9938	0.0124	0.9876
2.07	0.0192	0.9808	0.0385	0.9615	2.51	0.0060	0.9940	0.0121	0.9879
2.08	0.0188	0.9812	0.0375	0.9625	2.52	0.0059	0.9941	0.0117	0.9883
2.09	0.0183	0.9817	0.0366	0.9634	2.53	0.0057	0.9943	0.0114	0.9886
2.10	0.0179	0.9821	0.0357	0.9643	2.54	0.0055	0.9945	0.0111	0.9889
2.11	0.0174	0.9826	0.0349	0.9651	2.55	0.0054	0.9946	0.0108	0.9892
2.12	0.0170	0.9830	0.0340	0.9660	2.56	0.0052	0.9948	0.0105	0.9895
2.13	0.0166	0.9834	0.0332	0.9668	2.57	0.0051	0.9949	0.0102	0.9898
2.14	0.0162	0.9838	0.0324	0.9676	2.576	0.005	0.995	0.01	0.99
2.15	0.0158	0.9842	0.0316	0.9684	2.58	0.0049	0.9951	0.0099	0.9901
2.16	0.0154	0.9846	0.0308	0.9692	2.59	0.0048	0.9952	0.0096	0.9904
2.17	0.0150	0.9850	0.0300	0.9700	2.60	0.0047	0.9953	0.0093	0.9907
2.18	0.0146	0.9854	0.0293	0.9707	2.61	0.0045	0.9955	0.0091	0.9909
2.19	0.0143	0.9857	0.0285	0.9715	2.62	0.0044	0.9956	0.0088	0.9912

Table A.2 *(continued)*

z	One tail π beyond	One tail π remainder	Two tail π beyond	Two tail π remainder	z	One tail π beyond	One tail π remainder	Two tail π beyond	Two tail π remainder
2.63	0.0043	0.9957	0.0085	0.9915	3.25	0.0006	0.9994	0.0012	0.9986
2.64	0.0041	0.9959	0.0083	0.9917	3.291	0.0005	0.9995	0.001	0.999
2.65	0.0040	0.9960	0.0080	0.9920	3.30	0.0005	0.9995	0.0010	0.9990
2.70	0.0035	0.9965	0.0069	0.9931	3.35	0.0004	0.9996	0.0008	0.9992
2.75	0.0030	0.9970	0.0060	0.9940	3.40	0.0003	0.9997	0.0007	0.9993
2.80	0.0026	0.9974	0.0051	0.9949	3.45	0.0003	0.9997	0.0006	0.9994
2.85	0.0022	0.9978	0.0044	0.9956	3.50	0.0002	0.9998	0.0005	0.9995
2.90	0.0019	0.9981	0.0037	0.9963	3.55	0.0002	0.9998	0.0004	0.9996
2.95	0.0016	0.9984	0.0032	0.9968	3.60	0.0002	0.9998	0.0003	0.9997
3.00	0.0013	0.9987	0.0027	0.9973	3.65	0.0001	0.9999	0.0003	0.9997
3.05	0.0011	0.9989	0.0023	0.9977	3.719	0.0001	0.9999	0.0002	0.9998
3.090	0.001	0.999	0.002	0.998	3.80	0.0001	0.9999	0.0001	0.9999
3.10	0.0010	0.9990	0.0019	0.9981	3.891	0.00005	0.99995	0.0001	0.9999
3.15	0.0008	0.9992	0.0016	0.9984	4.000	0.00003	0.99997	0.00006	0.99994
3.20	0.0007	0.9993	0.0014	0.9988	4.265	0.00001	0.99999	0.00002	0.99998

From *Biometrika Tables for Statisticians*, Vol. 1, 3rd ed., by E.S. Pearson and H.O. Hartley, 1966, London: Cambridge University Press. Adapted with permission of the Biometrika Trustees.

Table A.3   Critical Values of Correlation Coefficients

df = N − 2	Level of significance for one-tailed test .05	.025	.01	.005	.0005	df = N − 2	Level of significance for one-tailed test .05	.025	.01	.005	.0005
	Level of significance for two-tailed test .10	.05	.02	.01	.001		Level of significance for two-tailed test .10	.05	.02	.01	.001
1	.9877	.9969	.9995	.9999	1.000	16	.4000	.4683	.5425	.5897	.7084
2	.9000	.9500	.9800	.9900	.9990	17	.3887	.4555	.5285	.5751	.6932
3	.8054	.8783	.9343	.9587	.9912	18	.3783	.4438	.5155	.5614	.6787
4	.7293	.8114	.8822	.9172	.9741	19	.3687	.4329	.5034	.5487	.6652
5	.6694	.7545	.8329	.8745	.9507	20	.3598	.4227	.4921	.5368	.6524
6	.6215	.7067	.7887	.8343	.9249	25	.3233	.3809	.4451	.4869	.5974
7	.5822	.6664	.7498	.7977	.8982	30	.2960	.3494	.4093	.4487	.5541
8	.5494	.6319	.7155	.7646	.8721	35	.2746	.3246	.3810	.4182	.5189
9	.5214	.6021	.6851	.7348	.8471	40	.2573	.3044	.3578	.3932	.4896
10	.4973	.5760	.6581	.7079	.8233	45	.2428	.2875	.3384	.3721	.4648
11	.4762	.5529	.6339	.6835	.8010	50	.2306	.2732	.3218	.3541	.4433
12	.4575	.5324	.6120	.6614	.7800	60	.2108	.2500	.2948	.3248	.4078
13	.4409	.5139	.5923	.6411	.7603	70	.1954	.2319	.2737	.3017	.3799
14	.4259	.4973	.5742	.6226	.7420	80	.1829	.2172	.2565	.2830	.3568
15	.4124	.4821	.5577	.6055	.7246	90	.1726	.2050	.2422	.2673	.3375
						100	.1638	.1946	.2301	.2540	.3211

Table A.3 is taken from Table VII of Fisher & Yates, *Statistical Tables for Biological, Agricultural and Medical Research* published by Longman Group Ltd., 1974.

Table A.4 Transformation of r to Z_r

r	z_r	r	z_r	r	z_r	r	z_r	r	z_r
.000	.000	.200	.203	.400	.424	.600	.693	.800	1.099
.005	.005	.205	.208	.405	.430	.605	.701	.805	1.113
.010	.010	.210	.213	.410	.436	.610	.709	.810	1.127
.015	.015	.215	.218	.415	.442	.615	.717	.815	1.142
.020	.020	.220	.224	.420	.448	.620	.725	.820	1.157
.025	.025	.225	.229	.425	.454	.625	.733	.825	1.172
.030	.030	.230	.234	.430	.460	.630	.741	.830	1.188
.035	.035	.235	.239	.435	.466	.635	.750	.835	1.204
.040	.040	.240	.245	.440	.472	.640	.758	.840	1.221
.045	.045	.245	.250	.445	.478	.645	.767	.845	1.238
.050	.050	.250	.255	.450	.485	.650	.775	.850	1.256
.055	.055	.255	.261	.455	.491	.655	.784	.855	1.274
.060	.060	.260	.266	.460	.497	.660	.793	.860	1.293
.065	.065	.265	.271	.465	.504	.665	.802	.865	1.313
.070	.070	.270	.277	.470	.510	.670	.811	.870	1.333
.075	.075	.275	.282	.475	.517	.675	.720	.875	1.354
.080	.080	.280	.288	.480	.523	.680	.829	.880	1.376
.085	.085	.285	.293	.485	.530	.685	.838	.885	1.398
.090	.090	.290	.299	.490	.536	.690	.848	.890	1.422
.095	.095	.295	.304	.495	.543	.695	.858	.895	1.447
.100	.100	.300	.310	.500	.549	.700	.867	.900	1.472
.105	.105	.305	.315	.505	.556	.705	.877	.905	1.499
.110	.110	.310	.321	.510	.563	.710	.887	.910	1.528
.115	.116	.315	.326	.515	.570	.715	.897	.915	1.557
.120	.121	.320	.332	.520	.576	.720	.908	.920	1.589
.125	.126	.425	.337	.525	.583	.725	.918	.925	1.623
.130	.131	.330	.343	.530	.590	.730	.929	.930	1.658
.135	.136	.335	.348	.535	.597	.735	.940	.935	1.697
.140	.141	.340	.354	.540	.604	.740	.950	.940	1.738
.145	.146	.345	.360	.545	.611	.745	.962	.945	1.783
.150	.151	.350	.365	.550	.618	.750	.973	.950	1.832
.155	.156	.355	.371	.555	.626	.755	.984	.955	1.886
.60	.161	.360	.377	.560	.633	.760	.996	.960	1.946
.165	.167	.365	.383	.565	.640	.765	1.008	.965	2.014
.170	.172	.370	.388	.570	.648	.770	1.020	.970	2.092
.175	.177	.375	.394	.575	.655	.775	1.033	.975	2.185
.180	.182	.380	.400	.580	.662	.780	1.045	.980	2.298
.185	.187	.385	.406	.585	.670	.785	1.058	.985	2.443
.190	.192	.390	.412	.590	.678	.790	1.071	.990	2.647
.195	.198	.395	.418	.595	.685	.795	1.085	.995	2.994

Table A.5   Critical Values of *t*

df	Level of significance for one-tailed test					
	.10	.05	.025	.01	.005	.0005
	Level of significance for two-tailed test					
	.20	.10	.05	.02	.01	.001
1	3.078	6.314	12.706	31.821	63.657	636.619
2	1.886	2.920	4.303	6.965	9.925	31.598
3	1.638	2.353	3.182	4.541	5.841	12.941
4	1.533	2.132	2.776	3.747	4.604	8.610
5	1.476	2.015	2.571	3.365	4.032	6.859
6	1.440	1.943	2.447	3.143	3.707	5.959
7	1.415	1.895	2.365	2.998	3.499	5.405
8	1.397	1.860	2.306	2.896	3.355	5.041
9	1.383	1.833	2.262	2.821	3.250	4.781
10	1.372	1.812	2.228	2.764	3.169	4.587
11	1.363	1.796	2.201	2.718	3.106	4.437
12	1.356	1.782	2.179	2.681	3.055	4.318
13	1.350	1.771	2.160	2.650	3.012	4.221
14	1.345	1.761	2.145	2.624	2.977	4.140
15	1.341	1.753	2.131	2.602	2.947	4.073
16	1.337	1.746	2.120	2.583	2.921	4.015
17	1.333	1.740	2.110	2.567	2.898	3.965
18	1.330	1.734	2.101	2.552	2.878	3.922
19	1.328	1.729	2.093	2.539	2.861	3.883
20	1.325	1.725	2.086	2.528	2.845	3.850
21	1.323	1.721	2.080	2.518	2.831	3.819
22	1.321	1.717	2.074	2.508	2.819	3.792
23	1.319	1.714	2.069	2.500	2.807	3.767
24	1.318	1.711	2.064	2.492	2.797	3.745
25	1.316	1.708	2.060	2.485	2.787	3.725
26	1.315	1.706	2.056	2.479	2.779	3.707
27	1.314	1.703	2.052	2.473	2.771	3.690
28	1.313	1.701	2.048	2.467	2.763	3.674
29	1.311	1.699	2.045	2.462	2.756	3.659
30	1.310	1.697	2.042	2.457	2.750	3.646
40	1.303	1.684	2.021	2.423	2.704	3.551
60	1.296	1.671	2.000	2.390	2.660	3.460
120	1.289	1.658	1.980	2.358	2.617	3.373
∞	1.282	1.645	1.960	2.326	2.576	3.291

Table A.5 is taken from Table III of Fisher & Yates; *Statistical Tables for Biological, Agricultural and Medical Research* published by Longman Group UK Ltd., 1974.

Table A.6 Critical Values of F

n, degrees of freedom (for greater mean square)

n_r	1	2	3	4	5	6	7	8	9	10	11	12	14	16	20	24	30	40	50	75	100	200	500	∞
1	161	200	216	225	230	234	237	239	241	242	243	244	245	246	248	249	250	251	252	253	253	254	254	254
	4,052	4,999	5,403	5,625	5,764	5,859	5,928	5,981	6,022	6,056	6,082	6,106	6,142	6,169	6,208	6,234	6,258	6,286	6,302	6,323	6,334	6,352	6,361	6,366
2	18.51	19.00	19.16	19.25	19.30	19.33	19.36	19.37	19.38	19.39	19.40	19.41	19.42	19.43	19.44	19.45	19.46	19.47	19.47	19.48	19.49	19.49	19.50	19.50
	98.49	99.00	99.17	99.25	99.30	99.33	99.34	99.36	99.38	99.40	99.41	99.42	99.43	99.44	99.45	99.46	99.47	99.48	99.48	99.49	99.49	99.49	99.50	99.50
3	10.13	9.55	9.28	9.12	9.01	8.94	8.88	8.84	8.81	8.78	8.76	8.74	8.71	8.69	8.66	8.64	8.62	8.60	8.58	8.57	8.56	8.54	8.54	8.53
	34.12	30.82	29.46	28.71	28.24	27.91	27.67	27.49	27.34	27.23	27.13	27.05	26.92	26.83	26.69	26.60	26.50	26.41	26.35	26.27	26.23	26.18	26.14	26.12
4	7.71	6.94	6.49	6.39	6.26	6.16	6.09	6.04	6.00	5.96	5.93	5.91	5.87	5.84	5.80	5.77	5.74	5.71	5.70	5.68	5.66	5.65	5.64	5.63
	21.20	18.00	16.69	15.98	15.52	15.21	14.98	14.80	14.66	14.54	14.45	14.37	14.24	14.15	14.02	13.93	13.83	13.74	13.69	13.61	13.57	13.52	13.48	13.46
5	6.61	5.79	5.41	5.19	5.05	4.95	4.88	4.82	4.78	4.74	4.70	4.68	4.64	4.60	4.56	4.53	4.50	4.46	4.44	4.42	4.40	4.38	4.37	4.36
	16.26	13.27	12.06	11.39	10.97	10.67	10.45	10.27	10.15	10.05	9.96	9.89	9.77	9.68	9.55	9.47	9.38	9.29	9.24	9.17	9.13	9.07	9.04	9.02
6	5.99	5.14	4.76	4.53	4.39	4.28	4.21	4.15	4.10	4.06	4.03	4.00	3.96	3.92	3.87	3.84	3.81	3.77	3.75	3.72	3.71	3.69	3.68	3.67
	13.74	10.92	9.78	9.15	8.75	8.47	8.26	8.10	7.98	7.87	7.79	7.72	7.60	7.52	7.39	7.31	7.23	7.14	7.09	7.02	6.99	6.94	6.90	6.88
7	5.59	4.74	4.35	4.12	3.97	3.87	3.79	3.73	3.68	3.63	3.60	3.57	3.52	3.49	3.44	3.41	3.38	3.34	3.32	3.29	3.28	3.25	3.24	3.23
	12.25	9.55	8.45	7.85	7.46	7.19	7.00	6.84	6.71	6.62	6.54	6.47	6.35	6.27	6.15	6.07	5.98	5.90	5.85	5.78	5.75	5.70	5.67	5.65
8	5.32	4.46	4.07	3.84	3.69	3.58	3.50	3.44	3.39	3.34	3.31	3.28	3.23	3.20	3.15	3.12	3.08	3.05	3.03	3.00	2.98	2.96	2.94	2.93
	11.26	8.65	7.59	7.01	6.63	6.37	6.19	6.03	5.91	5.82	5.74	5.67	5.56	5.48	5.36	5.28	5.20	5.11	5.06	5.00	4.96	4.91	4.88	4.86
9	5.12	4.26	3.86	3.63	3.48	3.37	3.29	3.23	3.18	3.13	3.10	3.07	3.02	2.98	2.93	2.90	2.86	2.82	2.80	2.77	2.76	2.73	2.72	2.71
	10.56	8.02	6.99	6.42	6.06	5.80	5.62	5.47	5.35	5.26	5.18	5.11	5.00	4.92	4.80	4.73	4.64	4.56	4.51	4.45	4.41	4.36	4.33	4.31
10	4.96	4.10	3.71	3.48	3.33	3.22	3.14	3.07	3.02	2.97	2.94	2.91	2.86	2.82	2.77	2.74	2.70	2.67	2.64	2.61	2.59	2.56	2.55	2.54
	10.04	7.56	6.55	5.99	5.64	5.39	5.21	5.06	4.95	4.85	4.78	4.71	4.60	4.52	4.41	4.33	4.25	4.17	4.12	4.05	4.01	3.96	3.93	3.91
11	4.84	3.98	3.59	3.36	3.20	3.09	3.01	2.95	2.90	2.86	2.82	2.79	2.74	2.70	2.65	2.61	2.57	2.53	2.50	2.47	2.45	2.42	2.41	2.40
	9.65	7.20	6.22	5.67	5.32	5.07	4.88	4.74	4.63	4.54	4.46	4.40	4.29	4.21	4.10	4.02	3.94	3.86	3.80	3.74	3.70	3.66	3.62	3.60
12	4.75	3.88	3.49	3.26	3.11	3.00	2.92	2.85	2.80	2.76	2.72	2.69	2.64	2.60	2.54	2.50	2.46	2.42	2.40	2.36	2.35	2.32	2.31	2.30
	9.33	6.93	5.95	5.41	5.06	4.82	4.65	4.50	4.39	4.30	4.22	4.16	4.05	3.98	3.86	3.78	3.70	3.61	3.56	3.49	3.46	3.41	3.38	3.36
13	4.67	3.80	3.41	3.18	3.02	2.92	2.84	2.77	2.72	2.67	2.63	2.60	2.55	2.51	2.46	2.42	2.38	2.34	2.32	2.28	2.26	2.24	2.22	2.21
	9.07	6.70	5.74	5.20	4.86	4.62	4.44	4.30	4.19	4.10	4.02	3.96	3.85	3.78	3.67	3.59	3.51	3.42	3.37	3.30	3.27	3.21	3.18	3.16
14	4.60	3.74	3.34	3.11	2.96	2.85	2.77	2.70	2.65	2.60	2.56	2.53	2.48	2.44	2.39	2.35	2.31	2.27	2.24	2.21	2.19	2.16	2.14	2.13
	8.86	6.51	5.56	5.03	4.69	4.46	4.28	4.14	4.03	3.94	3.86	3.80	3.70	3.62	3.51	3.43	3.34	3.26	3.21	3.14	3.11	3.06	3.02	3.00
15	4.54	3.68	3.29	3.06	2.90	2.79	2.70	2.64	2.59	2.55	2.51	2.48	2.43	2.39	2.33	2.29	2.25	2.21	2.18	2.15	2.12	2.10	2.08	2.07
	8.68	6.36	5.42	4.89	4.56	4.32	4.14	4.00	3.89	3.80	3.73	3.67	3.56	3.48	3.36	3.29	3.20	3.12	3.07	3.00	2.97	2.92	2.89	2.87
16	4.49	3.63	3.24	3.01	2.85	2.74	2.66	2.59	2.54	2.49	2.45	2.42	2.37	2.33	2.28	2.24	2.20	2.16	2.13	2.09	2.07	2.04	2.02	2.01
	8.53	6.23	5.29	4.77	4.44	4.20	4.03	3.89	3.78	3.69	3.61	3.55	3.45	3.37	3.25	3.18	3.10	3.01	2.96	2.89	2.86	2.80	2.77	2.75
17	4.45	3.59	3.20	2.96	2.81	2.70	2.62	2.55	2.50	2.45	2.41	2.38	2.33	2.29	2.23	2.19	2.15	2.11	2.08	2.04	2.02	1.99	1.97	1.96
	8.40	6.11	5.18	4.67	4.34	4.10	3.93	3.79	3.68	3.59	3.52	3.45	3.35	3.27	3.16	3.08	3.00	2.92	2.86	2.79	2.76	2.70	2.67	2.65
18	4.41	3.55	3.16	2.93	2.77	2.66	2.58	2.51	2.46	2.41	2.37	2.34	2.29	2.25	2.19	2.15	2.11	2.07	2.04	2.00	1.98	1.95	1.93	1.92
	8.28	6.01	5.09	4.58	4.25	4.01	3.85	3.71	3.60	3.51	3.44	3.37	3.27	3.19	3.07	3.00	2.91	2.83	2.78	2.71	2.68	2.62	2.59	2.57

df																								
19	1.88	1.90	1.91	1.94	1.96	2.00	2.02	2.07	2.11	2.15	2.21	2.26	2.31	2.34	2.38	2.43	2.48	2.55	2.63	2.74	2.90	3.13	3.52	4.38
	2.49	2.51	2.54	2.60	2.63	2.70	2.76	2.84	2.92	3.00	3.12	3.19	3.30	3.36	3.43	3.52	3.63	3.77	3.94	4.17	4.50	5.01	5.93	8.18
20	1.84	1.85	1.87	1.90	1.92	1.96	1.99	2.04	2.08	2.12	2.18	2.23	2.28	2.31	2.35	2.40	2.45	2.52	2.60	2.71	2.87	3.10	3.49	4.35
	2.42	2.44	2.47	2.53	2.56	2.63	2.69	2.77	2.86	2.94	3.05	3.13	3.23	3.30	3.37	3.45	3.56	3.71	3.87	4.10	4.43	4.94	5.85	8.10
21	1.81	1.82	1.84	1.87	1.89	1.93	1.96	2.00	2.05	2.09	2.15	2.20	2.25	2.28	2.32	2.37	2.42	2.49	2.57	2.68	2.84	3.07	3.47	4.32
	2.36	2.38	2.42	2.47	2.51	2.58	2.63	2.72	2.80	2.88	2.99	23.07	3.17	3.24	3.31	3.40	3.51	3.65	3.81	4.04	4.37	4.87	5.78	8.02
22	1.78	1.80	1.81	1.84	1.87	1.91	1.93	1.98	2.03	2.07	2.13	2.18	2.23	2.26	2.30	2.35	2.40	2.47	2.55	2.66	2.82	3.05	3.44	4.30
	2.31	2.33	2.37	2.42	2.46	2.53	2.58	2.67	2.75	2.83	2.94	3.02	3.12	3.18	3.26	3.35	3.45	3.59	3.76	3.99	4.31	4.82	5.72	7.94
23	1.76	1.77	1.79	1.82	1.84	1.88	1.91	1.96	2.00	2.04	2.10	2.14	2.20	2.24	2.28	2.32	2.38	2.45	2.53	2.64	2.80	3.03	3.42	4.28
	2.26	2.28	2.32	2.37	2.41	2.48	2.53	2.62	2.70	2.78	2.89	2.97	3.07	3.14	3.21	3.30	3.41	3.54	3.71	3.94	4.26	4.76	5.66	7.88
24	1.73	1.74	1.76	1.80	1.82	1.86	1.89	1.94	1.98	2.02	2.09	2.13	2.18	2.22	2.26	2.30	2.36	2.43	2.51	2.62	2.78	3.01	3.40	4.26
	2.21	2.23	2.27	2.33	2.36	2.44	2.449	2.58	2.66	2.74	2.85	2.93	3.03	3.09	3.17	3.25	3.36	3.50	3.67	3.90	4.22	4.72	5.61	7.82
25	1.71	1.72	1.74	1.77	1.80	1.84	1.87	1.92	1.96	2.00	2.06	2.11	2.16	2.20	2.24	2.28	2.34	2.41	2.49	2.60	2.76	2.99	3.38	4.24
	2.17	2.19	2.23	2.29	2.32	2.40	2.45	2.54	2.62	2.70	2.81	2.89	2.99	3.05	3.13	3.21	3.32	3.46	3.63	3.86	4.18	4.68	5.57	7.77
26	1.69	1.70	1.72	1.76	1.78	1.82	1.85	1.90	1.95	1.99	2.05	2.10	2.15	2.18	2.22	2.27	2.32	2.39	2.47	2.59	2.74	2.98	3.37	4.22
	2.13	2.15	2.19	2.25	2.28	2.36	2.42	2.50	2.58	2.66	2.77	2.86	2.96	3.02	3.09	3.17	3.29	3.42	3.59	3.82	4.14	4.64	5.53	7.72
27	1.67	1.68	1.71	1.74	1.76	1.80	1.84	1.88	1.93	1.97	2.03	2.08	2.13	2.16	2.20	2.25	2.30	2.37	2.46	2.57	2.73	2.96	3.35	4.21
	2.10	2.12	2.16	2.21	2.25	2.33	2.38	2.47	2.55	2.63	2.74	2.83	2.93	2.98	3.06	3.14	3.26	3.39	3.56	3.79	4.11	4.60	5.49	7.68
28	1.65	1.67	1.69	1.72	1.75	1.78	1.81	1.87	1.91	1.96	2.02	2.06	2.12	2.15	2.19	2.24	2.29	2.36	2.44	2.56	2.71	2.95	3.34	4.20
	2.06	2.09	2.13	2.18	2.22	2.30	2.35	2.44	2.52	2.60	2.71	2.80	2.90	2.95	3.03	3.11	3.23	3.36	3.53	3.76	4.07	4.57	5.45	7.64
29	1.64	1.65	1.68	1.71	1.73	1.77	1.80	1.85	1.90	1.94	2.00	2.05	2.10	2.14	2.18	2.22	2.28	2.35	2.43	2.54	2.70	2.93	3.33	4.18
	2.03	2.06	2.10	2.15	2.19	2.27	2.32	2.41	2.49	2.57	2.68	2.77	2.87	2.92	3.00	3.08	3.20	3.33	3.50	3.73	4.04	4.54	5.42	7.60
30	1.62	1.64	1.66	1.69	1.72	1.76	1.79	1.84	1.89	1.93	1.99	2.04	2.09	2.12	2.16	2.21	2.27	2.34	2.42	2.53	2.69	2.92	3.32	4.17
	2.01	2.03	2.07	2.13	2.16	2.24	2.29	2.38	2.47	2.55	2.66	2.74	2.84	2.90	2.98	3.06	3.17	3.30	3.47	3.70	4.02	4.51	5.39	7.56
32	1.59	1.61	1.64	1.67	1.69	1.74	1.76	1.82	1.86	1.91	1.97	2.02	2.07	2.10	2.14	2.19	2.25	2.32	2.40	2.51	2.67	2.90	3.30	4.15
	1.96	1.98	2.02	2.08	2.12	2.20	2.25	2.34	2.42	2.51	2.62	2.70	2.80	2.86	2.94	3.02	3.13	3.27	3.43	3.66	3.97	4.46	5.34	7.50
34	1.57	1.59	1.61	1.64	1.67	1.71	1.74	1.80	1.84	1.89	1.95	2.00	2.05	2.08	2.12	2.17	2.23	2.30	2.38	2.49	2.65	2.88	3.28	4.13
	1.91	1.94	1.98	2.04	2.08	2.15	2.21	2.30	2.38	2.47	2.58	2.66	2.76	2.82	2.89	2.97	3.08	3.21	3.38	3.61	3.93	4.42	5.29	7.44
36	1.55	1.56	1.59	1.62	1.65	1.69	1.71	1.78	1.82	1.87	1.93	1.98	2.03	2.06	2.10	2.15	2.21	2.28	2.36	2.48	2.63	2.86	3.26	4.11
	1.87	1.90	1.94	2.00	2.04	2.12	2.17	2.26	2.35	2.43	2.54	2.62	2.72	2.78	2.86	2.94	3.04	3.18	3.35	3.58	3.89	4.38	5.25	7.39
38	1.53	1.54	1.57	1.60	1.63	1.67	1.71	1.76	1.81	1.85	1.92	1.96	2.02	2.05	2.09	2.14	2.19	2.26	2.35	2.46	2.62	2.85	3.25	4.10
	1.84	1.86	1.90	1.97	2.00	2.08	2.14	2.22	2.32	2.40	2.51	2.59	2.69	2.75	2.82	2.91	3.02	3.15	3.32	3.54	3.86	4.34	5.21	7.35
40	1.51	1.53	1.55	1.59	1.61	1.66	1.69	1.74	1.79	1.84	1.90	.195	2.00	2.04	2.07	2.12	2.18	2.25	2.34	2.45	2.61	2.84	3.23	4.08
	1.81	1.84	1.88	1.94	1.97	2.05	2.11	2.20	2.30	2.37	2.49	2.56	2.66	2.73	2.80	2.88	2.99	3.12	3.29	3.51	3.83	4.31	5.18	7.31
42	1.49	1.51	1.54	1.57	1.60	1.64	1.68	1.73	1.78	1.82	1.89	1.94	1.99	2.02	2.06	2.11	2.17	2.24	2.32	2.44	2.59	2.83	3.22	4.07
	1.78	1.80	1.85	1.91	1.94	2.02	2.08	2.17	2.26	2.35	2.46	2.54	2.64	2.70	2.77	2.86	2.96	3.10	3.26	3.49	3.80	4.29	5.15	7.27
44	1.48	1.50	1.52	1.56	1.58	1.63	1.66	1.72	1.76	1.81	1.88	1.92	1.98	2.01	2.05	2.10	2.16	2.23	2.31	2.43	2.58	2.82	3.21	4.06
	1.75	1.78	1.82	1.88	1.92	2.00	2.06	2.15	2.24	2.32	2.44	2.52	2.62	2.68	2.75	2.84	2.94	3.07	3.24	3.46	3.78	4.26	5.12	7.24

Table A.6 (continued)

Each cell shows: top line = $p < .05$; bottom line = $p < .01$.

n_2	1	2	3	4	5	6	7	8	9	10	11	12	14	16	20	24	30	40	50	75	100	200	500	∞
46	4.05/7.21	3.20/5.10	2.81/4.24	2.57/3.76	2.42/3.44	2.30/3.22	2.22/3.05	2.14/2.92	2.09/2.82	2.04/2.73	2.00/2.66	1.97/2.60	1.91/2.50	1.87/2.42	1.80/2.30	1.75/2.22	1.71/2.13	1.65/2.04	1.62/1.98	1.57/1.90	1.54/1.86	1.51/1.80	1.48/1.76	1.46/1.72
48	4.04/7.19	3.19/5.08	2.80/4.22	2.56/3.74	2.41/3.42	2.30/3.20	2.21/3.04	2.14/2.90	2.08/2.80	2.03/2.71	1.99/2.64	1.96/2.58	1.90/2.48	1.86/2.40	1.79/2.28	1.74/2.20	1.70/2.11	1.64/2.02	1.61/1.96	1.56/1.88	1.53/1.84	1.50/1.78	1.47/1.73	1.45/1.70
50	4.03/7.17	3.18/5.06	2.79/4.20	2.56/3.72	2.40/3.41	2.29/3.18	2.20/3.02	2.13/2.88	2.07/2.78	2.02/2.70	1.98/2.62	1.95/2.56	1.90/2.46	1.85/2.39	1.78/2.26	1.74/2.18	1.69/2.10	1.63/2.00	1.60/1.94	1.55/1.86	1.52/1.82	1.48/1.76	1.46/1.71	1.44/1.68
55	4.02/7.12	3.17/5.01	2.78/4.16	2.54/3.68	2.38/3.37	2.27/3.15	2.18/2.98	2.11/2.85	2.05/2.75	2.00/2.66	1.97/2.59	1.93/2.53	1.88/2.43	1.83/2.35	1.76/2.23	1.72/2.15	1.67/2.06	1.61/1.96	1.58/1.90	1.52/1.82	1.50/1.78	1.46/1.71	1.43/1.66	1.41/1.64
60	4.00/7.08	3.15/4.98	2.76/4.13	2.52/3.65	2.37/3.34	2.25/3.12	2.17/2.95	2.10/2.82	2.04/2.72	1.99/2.63	1.95/2.56	1.92/2.50	1.86/2.40	1.81/2.32	1.75/2.20	1.70/2.12	1.65/2.03	1.59/1.93	1.56/1.87	1.50/1.79	1.48/1.74	1.44/1.68	1.41/1.63	1.39/1.60
65	3.99/7.04	3.14/4.95	2.75/4.10	2.51/3.62	2.36/3.31	2.24/3.09	2.15/2.93	2.08/2.79	2.02/2.70	1.98/2.61	1.94/2.54	1.90/2.47	1.85/2.37	1.80/2.30	1.73/2.18	1.68/2.09	1.63/2.00	1.57/1.90	1.54/1.84	1.49/1.76	1.46/1.71	1.42/1.64	1.39/1.60	1.37/1.56
70	3.98/7.01	3.13/4.92	2.74/4.08	2.50/3.60	2.35/3.29	2.23/3.07	2.14/2.91	2.07/2.77	2.01/2.67	1.97/2.59	1.93/2.51	1.89/2.45	1.84/2.35	1.79/2.28	1.72/2.15	1.67/2.07	1.62/1.98	1.56/1.88	1.53/1.82	1.47/1.74	1.45/1.69	1.40/1.62	1.37/1.56	1.35/1.53
80	3.96/6.96	3.11/4.88	2.72/4.04	2.48/3.56	2.33/3.25	2.21/3.04	2.12/2.87	2.05/2.74	1.99/2.64	1.95/2.55	1.91/2.48	1.88/2.41	1.82/2.32	1.77/2.24	1.70/2.11	1.65/2.03	1.60/1.94	1.54/1.84	1.51/1.78	1.45/1.70	1.42/1.65	1.38/1.57	1.35/1.52	1.32/1.49
100	3.94/6.90	3.09/4.82	2.70/3.98	2.46/3.51	2.30/3.20	2.19/2.99	2.10/2.82	2.03/2.69	1.97/2.59	1.92/2.51	1.88/2.43	1.85/2.36	1.79/2.26	1.75/2.19	1.68/2.06	1.63/1.98	1.57/1.89	1.51/1.79	1.48/1.73	1.42/1.64	1.39/1.59	1.34/1.51	1.30/1.46	1.28/1.43
125	3.92/6.84	3.07/4.78	2.68/3.94	2.44/3.47	2.29/3.17	2.17/2.95	2.08/2.79	2.01/2.65	1.95/2.56	1.90/2.47	1.86/2.40	1.83/2.33	1.77/2.23	1.72/2.15	1.65/2.03	1.60/1.94	1.55/1.85	1.49/1.75	1.45/1.68	1.39/1.59	1.36/1.54	1.31/1.46	1.27/1.40	1.25/1.37
150	3.91/6.81	3.06/4.75	2.67/3.91	2.43/3.44	2.27/3.14	2.16/2.92	2.07/2.76	2.00/2.62	1.94/2.53	1.89/2.44	1.85/2.37	1.82/2.30	1.75/2.20	1.71/2.12	1.64/2.00	1.59/1.91	1.54/1.83	1.47/1.72	1.44/1.66	1.37/1.56	1.34/1.51	1.29/1.43	1.25/1.37	1.22/1.33
200	3.89/6.76	3.04/4.71	2.65/3.88	2.41/3.41	2.26/3.11	2.14/2.90	2.05/2.73	1.98/2.60	1.92/2.50	1.87/2.41	1.83/2.34	1.80/2.28	1.74/2.17	1.69/2.09	1.62/1.97	1.57/1.88	1.52/1.79	1.45/1.69	1.42/1.62	1.35/1.53	1.32/1.48	1.26/1.39	1.22/1.33	1.19/1.28
400	3.86/6.70	3.02/4.66	2.62/3.83	2.39/3.36	2.23/3.06	2.12/2.85	2.03/2.69	1.96/2.55	1.90/2.46	1.85/2.37	1.81/2.29	1.78/2.23	1.72/2.12	1.67/2.04	1.60/1.92	1.54/1.84	1.49/1.74	1.42/1.64	1.38/1.57	1.32/1.47	1.28/1.42	1.22/1.32	1.16/1.24	1.13/1.19
1000	3.85/6.66	3.00/4.62	2.61/3.80	2.38/3.34	2.22/3.04	2.10/2.82	2.02/2.66	1.95/2.53	1.89/2.43	1.84/2.34	1.80/2.26	1.76/2.20	1.70/2.09	1.65/2.01	1.58/1.89	1.53/1.81	1.47/1.71	1.41/1.61	1.36/1.54	1.30/1.44	1.26/1.38	1.19/1.28	1.13/1.19	1.08/1.11
∞	3.84/6.64	2.99/4.60	2.60/3.78	2.37/3.32	2.21/3.02	2.09/2.80	2.01/2.64	1.94/2.51	1.88/2.41	1.83/2.32	1.79/2.24	1.75/2.18	1.69/2.07	1.64/1.99	1.57/1.87	1.52/1.79	1.46/1.69	1.40/1.59	1.35/1.52	1.28/1.41	1.24/1.36	1.17/1.25	1.11/1.15	1.00/1.00

n_1 degrees of freedom (for greater mean square)

Each n_2 level; $p < .05$ – top line; $p < .01$ – bottom line

From *Statistical Methods*, 7th ed. (pp. 480–483), by G.W. Snedecor and W.G. Cochran, 1980, Ames, Iowa: The Iowa State University Press. Copyright 1980 by The Iowa State University Press. Reprinted with permission.

Table A.7 Critical Values of the Studentized Range Statistic

df for S_{w^2}	$1-\alpha$	k = number of means or steps between ordered means								
		2	3	4	5	6	7	8	9	10
1	.95	18.0	27.0	32.8	37.1	40.4	43.1	45.4	47.4	49.1
	.99	90.0	135	164	186	202	216	227	237	246
2	.95	6.09	8.3	9.8	10.9	11.7	12.4	13.0	13.5	14.0
	.99	14.0	19.0	22.3	24.7	26.6	28.2	29.5	30.7	31.7
3	.95	4.50	5.91	6.82	7.50	8.04	8.48	8.85	9.18	9.46
	.99	8.26	10.6	12.2	13.3	14.2	15.0	15.6	16.2	16.7
4	.95	3.93	5.04	5.76	6.29	6.71	7.05	7.35	7.60	7.83
	.99	6.51	8.12	9.17	9.96	10.6	11.1	11.5	11.9	12.3
5	.95	3.64	4.60	5.22	5.67	6.03	6.33	6.58	6.80	6.99
	.99	5.70	6.97	7.80	8.42	8.91	9.32	9.67	9.97	10.2
6	.95	3.46	4.34	4.90	5.31	5.63	5.89	6.12	6.32	6.49
	.99	5.24	6.33	7.03	7.56	7.97	8.32	8.61	8.87	9.10
7	.95	3.34	4.16	4.69	5.06	5.36	5.61	5.82	6.00	6.16
	.99	4.95	5.92	6.54	7.01	7.37	7.68	7.94	8.17	8.37
8	.95	3.26	4.04	4.53	4.89	5.17	5.40	5.60	5.77	5.92
	.99	4.74	5.63	6.20	6.63	6.96	7.24	7.47	7.68	7.87
9	.95	3.20	3.95	4.42	4.76	5.02	5.24	5.43	5.60	5.74
	.99	4.60	5.43	5.96	6.35	6.66	6.91	7.13	8.32	7.49
10	.95	3.15	3.88	4 33	4.65	4.91	5.12	5.30	5.46	5.60
	.99	4.48	5.27	5.77	6.14	6.43	6.67	6.87	7.05	7.21
11	.95	3.11	3.82	4.26	4.57	4.82	5.03	5.20	5.35	5.49
	.99	4.39	5.14	5.62	5.97	6.25	6.48	6.67	6.84	6.99
12	.95	3.08	3.77	4.20	4.51	4.75	4.95	5.12	5.27	5.40
	.99	4.32	5.04	5.50	5.84	6.10	6.32	6.51	6.67	6.81
13	.95	3.06	3.73	4.15	4.45	4.69	4.88	5.05	5.19	5.32
	.99	4.26	4.96	5.40	5.73	5.98	6.19	6.37	6.53	6.67
14	.95	3.03	3.70	4.11	4.41	4.64	4.83	4.99	5.13	5.25
	.99	4.21	4.89	5.32	5.63	5.88	6.08	6.26	6.41	6.54
16	.95	3.00	3.65	4.05	4.33	4.56	4.74	4.90	5.03	5.15
	.99	4.13	4.78	5.19	5.49	5.72	5.92	6.08	6.22	6.35
18	.95	2.97	3.61	4.00	4.28	4.49	4.67	4.82	4.96	5.07
	.99	4.07	4.70	5.09	5.38	5.60	5.79	5.94	6.08	6.20
20	.95	2.95	3.58	3.96	4.23	4.45	4.62	4.77	4.90	5.01
	.99	4.02	4.64	5.02	5.29	5.51	5.69	5.84	5.97	6.09
24	.95	2.92	3.53	3.90	4.17	4.37	4.54	4.68	4.81	4.92
	.99	3.96	4.54	4.91	5.17	5.37	5.54	5.69	5.81	5.92
30	.95	2.89	3.49	3.84	4.10	4.30	4.46	4.60	4.72	4.83
	.99	3.89	4.45	4.80	5.05	5.24	5.40	5.54	5.56	5.76
40	.95	2.86	3.44	3.79	4.04	4.23	4.39	4.52	4.63	4.74
	.99	3.82	4.37	4.70	4.93	5.11	5.27	5.39	5.50	5.60
60	.95	2.83	3.40	3.74	3.98	4.16	4.31	4.44	4.55	4.65
	.99	3.76	4.28	4.60	4.82	4.99	5.13	5.25	5.36	5.45
120	.95	2.80	3.36	3.69	3.92	4.10	4.24	4.36	4.48	4.56
	.99	3.70	4.20	4.50	4.71	4.87	5.01	5.12	5.21	5.30
∞	.95	2.77	3.31	3.63	3.86	4.03	4.17	4.29	4.39	4.47
	.99	3.64	4.12	4.40	4.60	4.76	4.88	4.99	5.08	5.16

Table A.8   Critical Values of Chi–Square

df	Probability under H_0 that $\chi^2 \geq$ chi–square													
	.99	.98	.95	.90	.80	.70	.50	.30	.20	.10	.05	.02	.01	.001
1	.00016	.00063	.0039	.016	.064	.15	.46	1.07	1.64	2.71	3.84	5.41	6.64	10.83
2	.02	.04	.10	.21	.45	.71	1.39	2.41	3.22	4.60	5.99	7.82	9.21	13.82
3	.12	.18	.35	.58	1.00	1.42	2.37	3.66	4.64	6.25	7.82	9.84	11.34	16.27
4	.30	.43	.71	1.06	1.65	2.20	3.36	4.88	5.99	7.78	9.49	11.67	13.28	18.46
5	.55	.75	1.14	1.61	2.34	3.00	4.35	6.06	7.29	9.24	11.07	13.39	15.09	20.52
6	.87	1.13	1.64	2.20	3.07	3.83	5.35	7.23	8.56	10.64	12.59	15.03	.16.81	22.46
7	1.24	1.56	2.17	2.83	3.82	4.67	6.35	8.38	9.80	12.02	14.07	16.62	18.48	24.32
8	1.65	2.03	2.73	3.49	4.59	45.53	7.34	9.52	11.03	13.36	15.51	18.17	20.09	26.12
9	2.09	2.53	3.32	4.17	5.38	6.39	8.34	10.66	12.24	14.68	16.92	19.68	21.67	27.88
10	2.56	3.06	3.94	4.86	6.18	7.27	9.34	11.78	13.44	15.99	18.31	21.16	23.21	29.59
11	3.05	3.61	4.58	5.58	6.99	8.15	10.34	12.90	14.63	17.28	19.68	22.62	24.72	31.26
12	3.57	4.18	5.23	6.30	7.81	9.03	11.34	14.01	15.81	18.55	21.03	24.05	26.22	32.91
13	4.11	4.76	5.89	7.04	8.63	9.93	12.34	15.12	16.98	19.81	22.36	25.47	27.69	34.53
14	4.66	5.37	6.57	7.79	9.47	10.82	13.34	16.22	18.15	21.06	23.68	26.87	29.14	26.12
15	5.23	5.98	7.26	8.55	10.31	11.72	14.34	17.32	19.31	22.31	25.00	28.26	30.58	37.70
16	5.81	6.61	7.96	9.31	11.15	12.62	15.34	18.42	20.46	23.54	26.30	29.63	32.00	39.29
17	6.41	7.26	8.67	10.08	12.00	13.53	16.34	19.51	21.62	24.77	27.59	31.00	33.41	40.75
18	7.02	7.91	9.39	10.86	12.86	14.44	17.34	20.60	22.76	25.99	28.87	32.35	34.80	42.31
19	7.63	8.57	10.12	11.65	13.72	15.35	18.34	21.69	23.90	27.20	30.14	33.69	36.19	43.82
20	8.26	9.24	10.85	12.44	14.58	16.27	19.34	22.78	25.04	28.41	31.41	35.02	37.57	45.32
21	8.90	9.92	11.59	13.24	15.44	17.18	20.34	23.86	26.17	29.62	32.67	36.34	38.93	46.80
22	9.54	10.60	12.34	14.04	16.31	18.10	21.34	24.94	27.30	30.81	33.92	37.66	40.29	48.27
23	10.20	11.29	13.09	14.85	17.19	19.02	22.34	26.02	28.43	32.01	35.17	38.97	41.64	49.73
24	10.86	11.99	13.85	15.66	18.06	19.94	23.34	27.10	29.55	33.20	36.42	40.27	42.98	51.18
25	11.52	12.70	14.61	16.47	18.94	20.87	24.34	28.17	30.68	34.38	37.65	41.57	44.31	52.62
26	12.20	13.41	15.38	17.29	19.82	21.79	25.34	29.25	31.80	35.56	38.88	42.86	45.64	54.05
27	12.88	14.12	16.15	18.11	20.70	22.72	26.34	30.32	32.91	36.74	40.11	44.14	46.96	55.48
28	13.56	14.85	16.93	18.94	21.59	23.65	27.34	31.39	34.03	37.92	41.34	45.42	48.28	56.89
29	14.26	15.57	17.71	19.77	22.48	24.58	28.34	32.46	35.14	39.09	42.56	46.69	49.59	58.30
30	14.95	16.31	18.49	20.60	23.36	25.51	29.34	33.53	36.25	40.26	43.77	47.96	50.89	59.70

Table A.8 is taken from Table IV of Fisher & Yates; *Statistical Tables for Biological, Agricultural and Medical Research* published by Longman Group UK Ltd., 1974.

Table A.9   Values of Spearman r_s
for the .05 and .01 Levels of Significance

N	.05	.01	N	.05	.01
6	.886	–	19	.462	.608
7	.786	–	20	.450	.591
8	.738	.881	21	.438	.576
9	.683	.833	22	.428	.562
10	.648	.818	23	.418	.549
11	.623	.794	24	.409	.537
12	.591	.780	25	.400	.526
13	.566	.745	26	.392	.515
14	.545	.716	27	.385	.505
15	.525	.689	28	.377	.496
16	.507	.666	29	.370	.487
17	.490	.645	30	.364	.478
18	.476	.625			

Reprinted, by permission, from E.G. Olds, 1938, "Distributions of sums of squares of rank differences for small numbers of individuals," *Annals of Mathematical Statistics* 9: 133-148, and E.G. Olds, 1949, "The 5% significance levels for sums of squares of rank differences and a correction," *Annals of Mathematical Statistics* 20:117-118.

SUBJECT INDEX

2

24-hour recall, 178, 182, 183, 187, 230
2x2 factorial design, 259

A

abscissa. See axis
absolute zero, 79
abstracts
    published, 17, 19, 269
accuracy
    in data collection, 169, 177
acknowledgements, 61, 278, 281
alpha error, level, 95, 96, 135
alternative hypothesis, 94, 160, 208
analysis of covariance, 73, 151
animal subjects
    Welfare Act, 47, 68
ANOVA
    degrees of freedom, 112, 129, 131,
        137, 138, 140, 144, 157, 158
    F ratio, 128, 138, 139, 144
    factorial, 141, 142, 143, 144, 152, 210,
        211, 215, 253, 256, 259
    simple (one-way), 138
anthropometric assessment, 188
anxiety, 194, 289
apparatus, 1, 41, 49, 51, 54, 286
appendix
    statistical tables, 343
    survey questionnaires, 353
    thesis, 250
    web sites information, 343
applied research, 3, 10, 11, 43, 170, 206,
    246
approval page, 278, 280
article, 16, 18, 19, 20, 29, 37, 38, 39, 41,
    48, 60, 62, 128, 232, 249, 250, 254,
    265, 266, 267, 269, 270, 273, 274, 277,
    279, 283, 286, 291, 292, 293, 295, 304
assumptions, 6, 37, 39, 51, 86, 153, 155,
    162, 166, 241, 283, 313
attitudes, 3, 228, 243
audience for reports, 289, 290, 293
authority, 5, 6, 66

author's guidelines
    AJCN, 292
authorship, 60, 61, 62, 70
average, 75, 79, 82, 84, 91, 129, 176, 178,
    180, 182, 222, 237
avis effect, 202, 205, 215
axis, 88, 92, 254, 256, 262

B

background and significance, 299, 301,
    303, 307
background information, 17, 30, 32, 283,
    285
balance, 97, 100, 106, 187, 192, 204, 205,
    212, 245, 272
bar graphs, 254, 257
basal metabolic rate, 187
baseline, 315
basic research, 3, 43, 170, 207, 209, 295
behavioral research, 64, 71, 125, 168,
    193, 216
bell-shaped probability distribution, 86
Belmont report, 64, 71
bias
    interviewer, 180
biochemical index, 186
biochemical markers, 186, 191
biomedical research, 56, 65, 95, 308
BMDP statistical program, 314
body of thesis, 291

C

Campbell and Stanley notation system,
    197, 201, 202, 298
case-cohort studies, 229, 235, 237, 239,
    243
case-control studies, 56, 229, 235
catalog, 18, 25
categorical variables, 7, 39, 127
cathode ray tubes, 310
cause and effect, 4, 196, 215
central processing unit, 310
central tendency measures
    mean, 82
    median, 83

mode, 84
chapter style, 278
chemical assays, 53, 286
chi-square, 136, 156, 157, 158, 161, 162, 167, 251, 309, 323, 324, 325, 326
cholesterol, 7, 35, 36, 53, 79, 92, 93, 94, 97, 98, 107, 133, 134, 136, 142, 145, 148, 150, 151, 175, 197, 211, 232, 260, 262, 276
choosing the title, 27
cigarette smoking, 214
clinical assessment, 189
clinical trials, 40, 41, 55, 56, 101, 177, 217, 232
closed-ended questions, 222
co-authors, 60, 61
co-authorship, 60, 61
coding and consumption errors, 181
coefficient of determination, 118, 123
coefficient of non-determination, 118
coefficient of variation, 174
cohort studies, 229, 235, 237, 239, 243
computer(s)
    chips, 311
    floppy diskettes, 311
    hardware, 309, 310, 341
    megahertz, 311
    microsoft's DOS, 312
    microsoft's OS/2, 312
    microsoft's UNIX, 312
    microsoft's windows, 312
    operating system, 312, 341
conceptual definitions, 37
conclusions, 10, 21, 28, 44, 62, 74, 94, 250, 268, 269, 270, 272, 273, 277, 278, 279, 287, 289, 291, 293
concurrent validity, 175, 176
confidence
    interval, 118, 186, 254, 262, 266
    level, 99, 262
confidence in instruments
    practical consideration, 49
    reliability, 8, 171
    validity, 170
confidentiality, 65, 66, 67
consent, 46, 63, 64, 65, 66, 67, 69, 278, 279, 288
construct validity, 175, 176
content validity, 175, 176
contingency tables, 323
continuous variables, 36

control groups, trials, 204, 208, 209
controlling threats to external validity, 206
copyright, 14, 15, 62, 281
correlation
    coefficient, 103, 111, 112, 113, 114, 115, 117, 118, 121, 123
    formula, 111
    meaningfulness, 118, 123
    partial, 121
    Pearson, 327, 328
    significance, 112, 113, 115
    simple, 76, 107
    Spearman rank-order, 162, 164, 167, 170, 326, 327, 328, 329
correlation for prediction, 115
correlation studies, 233, 234, 243
creatinine, 188, 191, 194
criterion validity, 175, 177
critical evaluation, 23
crossover design, 205
cross-sectional studies, 237

D

data
    analysis, 50, 53, 127, 242, 243, 247, 289, 309, 313, 342
    collection, 169, 177, 195, 213, 220, 221, 231, 241, 242, 286, 289, 304, 310
data analysis, 73
data entry software, 312
Declaration of Helsinki, 64
deductive reasoning, 14, 16, 23, 300
definition
    conceptual, 37, 39
    operational, 37, 39, 44, 51
    research, 2
degrees of freedom, 112, 129, 131, 137, 138, 140, 144, 157, 158, 159, 161
delimitations, 1, 27, 37, 38, 39, 51
Delphi method, 195, 220, 228
dependent t test, 132, 133, 165, 167
describing analysis
    design, 50
    procedures, 53
descriptive research, 4, 195, 219, 220, 228, 232, 242
design. See research
developmental research, 4, 219

360

DEXA, 75
diet history
    intervention, 48
dietary assessment
    error sources, 180
    random errors, 173, 174, 178, 179,
        180, 181, 183, 190
    systematic errors, 178, 190
dietary history
    intake data, 190, 226
    questionnaire, 231
    records, 185, 186, 193
dietary intake methodology
    group dietary data, 229
    individual dietary methods, 230
    reproducibility, 173, 182
    validity of, 183
discrete variables, 36
discussion
    key compenents, 267
dissertation, 23, 29, 37, 39, 42, 63, 85,
    249, 250, 267, 269, 277, 278, 279, 281,
    282, 283, 284, 286, 288, 291, 293, 308
distribution
    kurtosis, 86, 89, 316, 317, 319
    normal, 73, 86, 87, 88, 89, 91, 92, 98,
        100, 174
    normal curve, 88, 89, 90, 91, 92
    skewness, 86, 88, 316, 317, 318, 319
documents, 18, 66, 192, 240, 242, 245
double-blind experiments, trials, 202
doubly-labeled water technique, 187
Duncan Multiple Range Test, 51, 137,
    140, 147, 153, 252, 253, 333, 334, 339
dyadic adjustment, 193

E

effect size, 86, 100, 196
electronic journal, 19
emic perspectives, 240
empirical method, 5, 6, 27
encyclopedias, 17
epidemiologic approaches to diet and
    disease, 181
epidemiologic descriptive research, 232
epidemiological research, 178, 186, 187,
    192, 195, 206, 235, 246
error
    type I, 95, 96, 97, 100
    type II, 95, 96, 100

error mean, variance, 50, 136, 138, 144,
    148, 173
estimating meaningfulness of treatments,
    131
ethical considerations
    copyright, 62, 281
    fabrication and falsification, 59
    guidelines for using animals in
        research proposal to NIH, 68
    guidelines for using human subjects in
        writing proposal, 67
    honorary authorship, 61
    informed consent, 46, 64, 65, 66
    IRB, 64, 65, 66, 67, 69
    misleading authorship, 60
    Nuremberg code, 63
    plagiarism, 58, 69
    protecting human subjects, 63
ethnography, 243, 245
etic perspectives, 240
ex post facto design, 213
excel, 313
exclusion criteria, 43, 304. See criteria
expected value, 120, 160, 162, 323
experimental mortality, treatment, 4, 41,
    165, 199, 200, 202, 205, 206, 207, 208,
    212, 213, 215, 286
experimental-type research, 17
    bias, 44, 81, 112, 129, 173, 174, 179,
        180, 183, 199, 201, 202, 203, 204,
        207, 208, 215, 227, 235, 238, 240,
        293
    pre-experimental design, 207, 215
    quasi-experimental design, 213, 215
    true experimental design, 207, 208,
        212, 215
    variations of experimental design, 195
external validity
    interaction of selection biases and
        experimental research, 203
    multiple-treatment interference, 203
    reactive effects of experimental
        arrangements, 203
    reactive or interactive effects of
        testing, 203

F

F table, ratio, test, 128, 136, 138, 139,
    140, 144, 153, 155, 166
fabrication, 58, 59, 69

face validity, 175
face-to-face contact, 220, 230
factor analysis, 309, 334, 336, 337, 339,
    340
factorial designs, 210, 211
falsification, 58, 60, 69
figures and illustrations
    preparation of, 21, 63, 249, 250, 254,
        259, 266, 272, 275, 291, 295
flat slope syndrome, 181
floppy diskettes, 311
follow-up analysis, 140, 206
food composition data and databases, 227,
    232
food disappearance data, 229
food frequency questionnaire
    by Block, 185, 232
    by Willett, 183, 232
    validity, 185
food records, 230, 231
formulating hypothesis, 7
framing problems, 27, 33, 300
Framingham heart study, 8, 11, 81
frequency analysis
    distribution, 174
    histogram, 256
Friedman's ANOVA, 166, 167
fundamental research, 168

G

gathering data, 8
generalizability, 8, 186, 197, 203, 204,
    217
graphic presentation of data, 254
graphical software, 312, 313
graphs
    data presentation, 254
    use, 254
Greenhouse-Geisser Epsilon, 337, 338,
    339, 341
group dietary data, 229
guidelines, 10, 63, 65, 67, 68, 69, 71, 191,
    229, 265, 274, 277, 278, 291, 292, 293,
    297, 301, 307

H

Halo effect, 201, 208, 208, 215
Hawthorne effect, 202, 203, 208, 215, 216
histograms, 250, 254, 257, 313, 320, 321

household food intake, 229
human subjects, 3, 4, 22, 43, 47, 63, 64,
    67, 70, 71, 213, 215, 269, 270, 286,
    305
Huynh-Feldt Epsilon, 337, 338, 339, 341
hypotheses

    alternative, 93
    experimenal-type research, 8
    experimental-type reseach, 300
    null, 36
    research, 9, 28, 37, 93
    statistical, 266
    testing, 93

I

identifying problem, 228
inclusion criteria, 42
independent t test, 130, 131, 135, 165,
    167, 209
Index Medicus, 17
indexes, 17, 18
inductive reasoning, 15, 16, 219, 300
inference, 44, 79, 80, 99
inferential statistics, 73, 79, 103
informed consent, 46, 64, 65, 66, 67, 69
institutional animal care and use
    committee, 47, 69
institutional review board, 46, 64, 65, 67,
    69, 70
instrumentation, 1, 21, 22, 41, 177, 199,
    205, 215, 285, 286, 305
intercept, straight-line regression, 331,
    332
internal consistency, criticism, 30, 171,
    292
internal validity
    Avis effect, 202, 205, 215
    blind, 205
    controlling threats, 204
    double blind, 205
    experimental mortality, 206
    Halo effect, 201, 205, 208, 215
    Hawthorne effect, 202, 203, 208, 215
    history, 198
    instrumentation, 206
    maturation, 198
    placebo, 205
    randomization, 204
    selection bias, 200

selection-maturation interaction, 201
testing, 199
uncontrolled threats, 205
internet, 17, 18, 25
interobserver, 171, 172
interpreting results, 1, 9, 10
intervention studies, 4, 48, 236, 237
interviewer bias, 180, 199
interviews, 219, 220, 222, 226, 230, 240, 241, 242, 243
intraobserver, 171, 172
intuition, 5
invalidity, 197
inverse associations, 102, 175

J

journal of nutrition, 292
journals
writing journal articles, 292

K

Kendall correlation coefficient, 162, 168
keywords, 18, 22, 270, 271
Kruskal-Wallis ANOVA, 166, 167, 170
kurtosis, 86, 89, 316, 317, 318, 319

L

laboratory animal care, 68
laboratory control, research, 3, 10
lead-in, introduction, 1, 13, 21, 31, 32, 33
length of questionnaires, 244
leptokurtic, 89
library information system, 18, 19
limitations, 1, 4, 6, 27, 37, 38, 39, 51, 152, 190, 231, 232, 234, 235, 267, 268, 278, 304, 306
line graph, 254
linear regression analysis, 123, 309, 330
literature, 1, 4, 13, 14, 16, 18, 19, 20, 21, 22, 23, 24, 25, 29, 32, 33, 36, 40, 60, 70, 219, 220, 238, 240, 243, 267, 272, 278, 282, 283, 285, 289, 291, 296, 300, 303, 304, 305
longitudinal studies, 4, 219
lotus 1-2-3, 313

M

main effects, 141, 143
Mann-Whitney-Wilcoxon, 165
manuscripts, 60, 265, 274, 288, 291, 292, 294
mathematical concepts, 82, 88, 98, 170, 187
maturation, 198, 201, 204, 208, 215
MAXICON principle, 50, 54, 173
mean, 44, 46, 48, 51, 75, 82, 84, 86, 87, 88, 89, 90, 91, 92, 93, 94, 96, 118, 120, 123, 128, 130, 131, 133
meaningfulness of statistics, 96, 100, 118, 123, 131
measurement
error, 173
interval, 170
nominal, 170
ordinal, 170
ratio, 170
reliability, 171
validity, 175
median, 82, 83, 86, 87, 88, 170, 327, 328
medline, 17, 20
megahertz, 311
menopause, 106, 125, 284
meta-analysis, 238
methods, 4, 5, 6, 8, 10, 11, 14, 22, 26, 27, 37, 38, 40, 41, 42, 46, 50, 53, 55, 56, 58, 63, 65, 67, 68, 74, 81, 94, 101, 102, 117, 140, 147, 154, 155, 166, 168, 171, 177, 178, 179, 180, 183, 184, 185, 186, 190, 191, 216, 217, 222, 229, 230, 231, 232, 241, 243, 244, 247, 248, 266, 270, 271, 274, 275, 277, 278, 282, 286, 289, 291, 293, 295, 297, 299, 301, 302, 304, 306, 307, 308, 314
Microsoft's DOS, OS/2, UNIX, LINX, 312
migrant studies, 234, 243
Minitab, 74
misconduct in science, 57, 58, 71
misleading authorship, 60
mode, 82,84, 86, 87, 88, 316, 318, 319
multiple comparison procedure, 335
multiple form, 171
multiple regression prediction equation, 121
multivariate analysis, 246

N

National Health and Nutrition
    Examination Surveys, 228
National Institutes of Health, 67, 68, 301,
    308
nationwide food consumption surveys, 230
naturalistic inquiry, framework, 15
negatively skewed curve, distribution, 88
Newman-Keuls test, 141, 147
nominal level, 78
non-parametric statistics, 73
non-parametric techniques
    chi-square, 136, 156, 157, 158, 161,
        162, 167, 251, 309, 323, 324, 325,
        326
    coefficient, 76, 103, 111, 112, 113,
        114, 115, 117, 118, 121, 123, 162,
        170, 171, 174, 177, 183, 260, 262,
        309, 324, 325, 326, 327, 328
    contingency, 159, 309, 323, 324, 325,
        326
    Friedman ANOVA, 166, 167
    Kruskal Wallis ANOVA, 166, 167,
        170
    Mann-Whitney-Wilcoxon test, 165
    Spearman rank correlation coefficient,
        162
    Wilcoxon matched-pair signed ranks
        test, 165
normal probability curve, 92
null hypothesis, 36, 86, 94, 95, 96, 100,
    127, 128, 135, 136, 157, 158, 159, 161,
    164, 196, 208, 330, 336
Nuremberg code, 63
nutrient intakes, 29, 178, 181, 182, 183,
    184, 190, 193, 199, 230
nutrition research, surveys, 3, 4, 38, 73,
    75, 191, 192, 206, 226, 228, 230, 243,
    245, 271
nutrition research, surveys, 101, 308

O

objectivity, 7, 8
observational research, 219
odds ratio, 185, 186
omega squared, 128, 132
one group pretest-posttest design, 210,
    215

one-tailed test, hypothesis, 153, 157, 161
one-way analysis of variance, 333
one-way chi-square, 156, 157, 162
open-ended questions, 222, 245, 247
operating systems, 312
operational definition, 37, 39, 51
oral presentation, 288, 289, 290, 293, 296
order effect, 246
ordinal level, scale, 78, 86, 165, 166, 170
osteoporosis, 8, 9, 22, 42, 43, 107, 268,
    284, 285

P

paired test, observations, 97, 111, 133,
    134, 165
parameter, 25, 45, 55, 82, 83, 97, 99, 153,
    155, 331, 332
parametric test, 155, 156, 165, 166, 168
partial correlation, 121
Pearson correlation coefficient, 327, 328
peer review, 293
pellagra, 177, 192, 233, 245
per capita intakes, 229
percentile, 170
personal interviews, 219, 242
photographs, 240, 242, 249, 250, 263,
    272, 287, 291
pie charts, 254, 313
pilot study, test, 180, 190
placebo, 201, 205, 215
plagiarism, 58, 69
planning the work
    presentations, 288
    writing, 22
platokurtic, 89
population, defined, 44
population-based study, 4, 28, 42, 44
portion size, 180, 181, 182, 184, 185, 191
positively skewed distribution, 88
poster presentation, 249, 269, 288, 290
power, 96
precision in dietary assessment
    in food records, 182
    of 24-hr recalls, 182
    of dietary histories, 182
    of food frequency questionnaires, 183
predicted score, 115
prediction
    equation for, 121, 186
    line of best fit, 115

standard error, 118, 120, 122, 123
predictive validity, 175, 177
predictor variables, 121
pre-experimental design, 207, 215
preliminary sources, research report, 17
presentations
    oral, 288, 290
    poster, 249
pre-testing, 213
pretest-posttest randomized groups
    design, 210, 215
prevalence, 31, 32, 34, 92, 237
primary references, 16
probability
    alpha, 94
    beta, 95
    level, 99, 129, 251
Probability
    distribution, 92
problem
    defining, 7
    delimiting, 7
problem solving, methods of
    scientific, 5, 7
    unscientific, 5
problem statement, 30, 33, 34, 39, 283
procedures, 1, 21, 41, 44, 47, 49, 51, 53,
    54, 65, 66, 67, 68, 69, 73, 74, 75, 80,
    94, 153, 175, 179, 188, 190, 195, 196,
    212, 216, 219, 220, 228, 270, 278, 286,
    304, 305, 314
productivity, 202, 305, 306
progress report, 301, 303, 306
proposal
    research, 307
proposal, strong, fail, 305, 306, 307
prospective studies, 235, 238
prospectus, 1
protocols, 56, 64, 69
publication, submitting the work for, 292
purposes of literature review, 16, 24, 285
p-value, 339

Q

quality assurance, 2
quality control, 66, 69, 173, 179, 189
quantile, 316, 317, 318, 319
quantitative research, 240
quasi-experimental design, 11, 213, 215
quatro Pro, 313

questionnaire, 29, 38, 155, 171, 172, 175,
    177, 178, 179, 181, 182, 183, 185, 186,
    190, 191, 192, 193, 195, 219, 220, 222,
    224, 225, 226, 227, 228, 231, 235, 242,
    244, 247, 248, 279, 288

R

random access memory, 311
randomization, 204, 205, 208, 209, 214,
    266
range, 18, 30, 31, 42, 49, 67, 79
range score, 85
rating, 155, 163, 194
ratio level, 79, 85
rationalistic method, 5, 6
raw data, 250, 278, 279, 288
reactive effects, 203, 207, 215
reasoning
    deductive, 14, 16, 23, 300
    inductive, 15, 219, 300
reliability, 8, 49, 73, 97, 101, 112, 169,
    170, 171, 172, 173, 177, 182, 187, 189,
    206, 216, 220, 246, 247, 272
repeated measures, 166, 167, 178, 210,
    309, 333, 336, 337, 338, 339, 340
repeated measures of analysis of variance,
    166, 178
repetition, 127, 225
reproducibility of dietary intake
    methodology studies, 173, 182
research
    analytical, 232
    applied, 3, 10, 43, 170, 206
    basic, 3, 10, 43, 170, 207, 209
    case-control, 105, 185, 229, 235, 243
    cohort, 221, 229, 235, 237, 239, 243
    controlled trials, 236
    cross-sectional, 219, 228, 237
    definition, 2
    descriptive, 4, 195, 219, 220, 228, 232,
        242
    design, 21, 66, 67, 97, 100, 141, 196,
        197, 205, 207, 213, 215, 232, 302,
        304
    experimental, 4, 10, 50, 81, 99, 127,
        135, 195, 196, 207, 215
    field, 3
    hypothesis, 9, 28, 37, 93
    laboratory, 3, 10
    literature, 21

longitudinal, 4, 199, 219
methods, 242, 243
migrant, 234, 243
pre-experimental, 195, 207, 215
proposal, 14, 65, 249, 299, 306
protocol, 64, 69, 299, 307
qualitative, 4, 50, 155, 195, 219, 239,
    241, 242, 272
special exposure group, 234
true experimental, 195, 207, 208, 212,
    215
residual variance, 152, 229
respondent bias, 180
response rate, 220, 222, 225, 244
*results, 1, 3, 4, 8, 10, 14, 20, 21, 23, 27,
    34, 36, 38, 39, 41, 54, 57, 58, 60, 61,
    62, 63, 74, 82, 94, 133, 134, 135, 146,
    150, 151, 153, 157, 173, 179, 182, 183,
    185, 197, 201, 203, 204, 211, 215, 226,
    227, 229, 249, 250, 252, 256, 266, 267,
    269, 271, 272, 277, 278, 282, 286, 287,
    289, 291, 293, 300, 302, 304, 305*
*review of literature, 1, 19, 23, 29, 278,
    282, 284, 285, 291*

S

sample mean, 82, 129, 130, 331
sampling
    distribution, 92, 137, 157
    error, 99
    procedures, 44, 73
    random, 81, 99, 207
    random assignment, 81, 99, 200, 204,
        208, 209, 213, 215
    stratified-random, 81, 99, 180, 230
    systematic, 81
scatter diagram, 260
Scheffe test, 137, 140, 147, 153
*science, 4, 11, 12, 18, 25, 40, 47, 51, 57,
    58, 62, 70, 74, 75, 101, 153, 168, 191,
    216, 219, 240, 275, 280, 295, 296, 297,
    306, 314*
scientific
    inquiry, 12
    method, 1, 5, 6, 9, 10, 11
scientific method of problem solving, 5, 7
*scores, 36, 49, 79, 82, 84, 85, 90, 93, 111,
    112, 113, 132, 135, 139, 154, 155, 165,
    171, 172, 176, 177, 200, 208, 212, 305*
secondary sources, 17

self-administered questionnaire, 221
*semi-quantitative food frequency
    questionnaire, 185*
significance
    of the study, 1, 27, 39, 283
    statistical, 154
significance level, 97, 186
skewed, negatively, 88
skewed, positively, 88, 137
skewness, 86, 88, 316, 317, 318, 319
*slope, straight-line regression, 115, 117,
    181, 262, 330, 331*
*software, 18, 20, 53, 74, 309, 311, 312,
    313, 314, 341*
*Solomon four group design, 205, 207, 211,
    212, 215*
**Spearman's correlation coefficient, 162,
    164, 167, 170, 326, 328, 329**
spearman's correlation coefficient, 327
special exposure group, 234, 243
specific aims, 299, 301, 302, 303, 304, 307
split-half technique, 171
standard
    deviation, 51, 82, 84, 85, 88, 89, 90,
        92, 97, 98, 121, 174, 190, 251, 254
    error of mean or estimate, 51, 118, 120,
        122
standard normal distribution, 86, 89, 98
*Stanley and Campbell notation system,
    197, 298, 201, 202*
STATA, 314
STATISTICA, 314
*statistical analysis, 9, 50, 75, 80, 96, 101,
    150, 179, 210, 286, 315, 342*
*statistical analysis system (SAS), 51, 53,
    74, 314, 331, 342*
statistical analytical system, 53
    applications, 74
    differences, 314
    SPSS, 51, 74, 314
*statistical package for the social science s
    (SPSS), 51, 74, 314*
stratified random, 81, 99
student's t test, 128
subjects
    characteristics, 42
    how many, 42
    protection, 46
*sum of squares, 138, 152, 332, 334, 335,
    337, 338, 341*
survey

Delphi, 228, 243
  normative, 195, 219, 220, 228, 243
  questionnaire, 29, 38, 155, 171, 172,
    175, 177, 178, 179, 180, 181, 182,
    183, 185, 186, 190, 195, 219, 220,
    222, 224, 225, 226, 228, 231, 235,
    242, 279, 288
systematic error, 178, 179, 181, 183, 190
systematic sampling, 81

T

t tests
  and power, 135
  dependent, 132, 165, 167
  independent, 130, 131, 135, 165, 167,
    209
  one-tailed, 94
  two-tailed, 94, 112
*tables, 21, 31, 32, 34, 102, 162, 227, 249,
  250, 266, 272, 275, 278, 282, 287, 291,
  323, 343*
tenacity, 5
ten-state nutrition survey, 75
*title, 18, 19, 20, 27, 28, 29, 39, 45, 47, 51,
  61, 67, 157, 200, 251, 253, 278, 279,
  280, 282, 290, 291, 305, 316, 320, 322,
  323, 327, 330, 332, 333, 335, 337, 340*
title page, 19, 278, 279, 280, 292
total variance, 132, 136
*treatments, 4, 8, 35, 41, 46, 48, 49, 66, 67,
  80, 95, 131, 135, 136, 142, 200, 204,
  205, 206, 207, 211, 215, 286*
triangulation, 242, 274, 276
true experimental design, 207, 212, 215
*true intake, variance, 50, 136, 138, 143,
  178*
truth table, 95
Tukey test, 51, 137, 255
two-tailed, hypothesis test, 94, 112
two-way chi-square
  contingency table, 159, 309, 323, 324,
    325
type I error, 95, 100
type II error, 95, 86, 97, 100

U

univariate analysis, model
  procedure, test, 316, 317, 319, 337,
    338, 340, 342

unscientific methods of problem solving, 5
*use of biochemical markers to validate
  dietary data, 186*

V

validity
  concurrent, 175, 176
  construct, 175, 176
  content, 175, 176
  criterion, 175
  external, 195, 197, 202, 203, 206, 215,
    220
  face, 175
  internal, 8, 197, 198, 202, 203, 204,
    206, 210, 213, 215
  predictive, 177
validity in dietary assessment methods
  24-hr recalls, 183
  of dietary history, 184, 191, 231
  of food frequency questionnaires, 185
  of food record, 184, 230, 231
variability measure
  standard deviatin, 91, 254
  standard deviation, 82, 84, 85, 88, 89,
    90, 92, 97, 98, 118, 120, 121, 122,
    128, 129, 131, 135, 136, 174, 175,
    179, 190, 251
variables
  categorical, 7, 39, 127
  control, 34
  dependent, 7, 27, 34, 35, 49, 73, 108,
    128, 136, 197, 206, 209, 215, 250,
    251, 252, 253, 254, 256, 259, 260,
    262, 263, 272, 300
  extraneous, 50, 196
  independent, 7, 50, 124, 137, 141, 142,
    143, 152, 209, 210, 214, 251, 252,
    254, 256, 259, 260, 262, 263
  moderator, 34
  treatment, 7, 197
variablity measure
  standard deviation, 84
variance
  error, 50, 136, 138, 144, 173
  true, 50, 136, 138, 143, 144

W

Wilcoxon matched pairs test, 165
within-person variation, 186

World Health Organization, 193, 246, 247
world wide web (WWW), 18
www browsing software packages, 18

Y

y-intercept, regression, 51, 73, 103, 105,
107, 108, 115, 117, 118, 120, 121, 122,
123, 124, 125, 151, 152, 162, 200, 204,

215, 260, 262, 309, 330, 331, 332

Z

z distribution, 90, 93
z scores, 90, 93